Marine Corps Aircraft: 1913-2000

by
Major John M. Elliott
U.S. Marine Corps (Retired)

Occasional Paper

HISTORY AND MUSEUMS DIVISION
HEADQUARTERS, U.S. MARINE CORPS
WASHINGTON, D.C.

2002

PCN 19000411600

Other Publications in the Occasional Papers Series

Vietnam Histories Workshop: Plenary Session. Jack Shulimson, editor. 9 May1983. 31 pp.

Vietnam Revisited; Conversation with William D. Broyles, Jr. Colonel John G. Miller, USMC, editor. 11 December 1984. 48 pp.

Bibiograpby on Khe Sanh USMC Participation. Commander Ray W. Strubbe, CHC, USNR (Ret), compiler. April 1985. 54 pp.

Alligators, Buffaloes, and Bushmasters: The History of the Development of the LVT Through World War II. Major Alfred Dunlop Bailey, USMC (Ret). 1986. 272 pp.

Leadership Lessons and Remembrances from Vietnam. Lieutenant General Herman Nickerson, Jr., USMC (Ret). 1988. 93 pp.

The Problems of U.S. Marine Corps Prisoners of War in Korea. James Angus MacDonald, Jr. 1988. 295 pp.

John Archer Lejeune, 1869-1942, Register of His Personal Papers. Lieutenant Colonel Merrill L. Bartlett, USMC (Ret). 1988. 123 pp.

To Wake Island and Beyond: Reminiscences. Brigadier General Woodrow M. Kessler, USMC (Ret). 1988. 145 pp.

Thomas Holcomb, 1879-1965, Register of His Personal Papers. Gibson B. Smith. 1988. 229 pp.

Curriculum Evolution, Marine Corps Command and Staff College, 1920-1988. Lieutenant Colonel Donald F. Bittner, USMCR. 1988. 112 pp.

Herringbone Cloak-GI Dagger, Marines of the OSS. Major Robert E. Mattingly, USMC. 1989. 315 pp.

The Journals of Marine Second Lieutenant Henry Bulls Watson, 1845-1848. Charles R. Smith, editor. 1990. 420 pp.

When the Russians Blinked: The U.S. Maritime Response to the Cuban Missile Crisis. Major John M. Young, USMCR. 1990. 246 pp.

Marines in the Mexican War. Gabrielle M. Neufeld Santelli. Edited by Charles R. Smith. 1991. 63 pp.

The Development of Amphibious Tactics in the U.S. Navy. General Holland M. Smith, USMC (Ret). 1992. 89 pp.

James Guthrie Harbord, 1866-1947, Register of His Personal Papers. Lieutenant Colonel Merrill L. Bartlett, USMCR. 1995. 47 pp.

The Impact of Project 100,000 on the Marine Corps. Captain David A. Dawson, USMC. 1995. 247 pp.

Occasional Papers

The History and Museums Division has undertaken the publication for limited distribution of various studies, theses, compilations, bibliographies, monographs, and memoirs, as well as proceedings at selected workshops, seminars, symposia, and similar colloquia, which it considers to be of significant value for audiences interested in Marine Corps history. These "Occasional Papers," which are chosen for their intrinsic worth, must reflect structured research, present a contribution to historical knowledge not readily available in published sources, and reflect original content on the part of the author, compiler, or editor. It is the intent of the division that these occasional papers be distributed to selected institutions, such as service schools, official Department of Defense historical agencies, and directly concerned Marine Corps organizations, so the information contained therein will be available for study and exploitation.

Foreword

Marine Corps aviation is the story of machines and men (and now women). Just as individuals are introduced in historical narratives by full name, rank, and service, so too are the various aircraft flown by Marines. History and Museums Division writing guides dating back to World War II have required aircraft to be identified by manufacturer, model number, and name using contemporary nomenclature. Getting this right can be a research project in itself. This is made more accurate and easier by this occasional paper that updates two previous division publications—*Marine Corps Aircraft, 1913-1960*, published in 1961 and subsequently revised in 1967. The author has reviewed and expanded the previous efforts into a comprehensive edition that lists Marine aircraft from the Curtis E-1 through the Lockheed KC-130J Hercules. The 470 entries provide the most complete reference for historians and curators to date.

Along with obvious scholarship, Major John M. "Jack" Elliott brings his own Marine Corps experience to bear on this effort. Born in Canada in 1923, Elliott subsequently moved with his family to Santa Barbara, California. He enlisted in the Marine Corps in 1942 and was sent to Aircraft Engineering Squadron 22 at Naval Air Station North Island, San Diego, California. Assigned as an aviation ordnance man, Elliott went overseas to Samoa and then to Efate, New Hebrides in 1943. The following year found him on Green Island and Emirau arming fighters and dive-bombers attacking Rabaul. Promoted to technical sergeant, he moved to the Philippines, returning to the United States in 1945. After squadron and instructional duties, Elliott was promoted to master sergeant and served in Korea with Marine Fighter Squadron 214 and 1st Marine Aircraft Wing. Commissioned in 1953, he worked on ordnance projects with the Naval Ordnance Plant, Indianapolis, Indiana; was assigned with Marine Aircraft Group 24, 1st Marine Aircraft Wing; and at various stations in the Far East; Quantico, Virginia, and Beaufort, South Carolina. In 1965, he was sent to Chu Lai in South Vietnam making Elliott a "three war" Marine. Promoted to major, he was the Marine Corps Development Center's ordnance officer until retiring in 1966.

Since retirement, Major Elliott has been active with the Smithsonian Institution, the Marine Corps Air-Ground Museum, the Naval Aviation Historical Office, and in updating this historical reference pamphlet among other projects.

J.W. RIPLEY
Colonel, U.S. Marine Corps (Retired)
Director of Marine Corps History and Museums

Introduction

For a number of years I have wanted to bring the Marine Corps Historical Reference Pamphlet "Marine Corps Aircraft 1913 – 1965" up to date. However, upon closer examination it was ascertained that there were numerous deficiencies in the original text that should be corrected. A number of aircraft were identified as being assigned to the Marine Corps, which in fact they never were. Some of this was due to information obtained in planning directives. In all cases where there was a question, a thorough search was made of the Aircraft History Cards for that type aircraft. There is a card for every aircraft the Navy has purchased since 1922 which are the official record of aircraft assignments. In this search it was found that there were a number of aircraft types that were not identified as being assigned to Marine Corps aviation.

Through the years, there have been numerous changes in the designation of a specific aircraft for a number of reasons. In each case these were picked up as a new aircraft. There were also a few cases where the aircraft designation had been changed prior to the assignment to the Marine Corps, but the old designation was used. In the case of engine identification a common entry was Wright Whirlwind or Pratt & Whitney Wasp Junior. This is similar to saying you have a Chevrolet or Buick but fails to completely identify what model you have.

In this study, I have attempted to correct these errors and fill in the missing data. This is not intended to be a criticism of the original authors but an attempt to make the pamphlet a more reliable reference tool in this highly complex subject.

Marine Corps Aircraft 1917 – 2000

All aircraft operated by the United States Marine Corps have been procured by the U.S. Navy and consequently have been designated by the same systems as established by the Navy. In July 1911, the first two U.S. Navy aircraft, the A-1 and A-2 were purchased from Glenn Curtiss. An Aviation Camp was established at Greenbury Point, Annapolis, Md. On 22 May 1912, 1stLt Alfred A. Cunningham was assigned to flight instruction there, which is regarded as the birthday of Marine Corps aviation. In the beginning, there was no official assignment of aircraft to the Marine Corps as there was no aviation organization within the Corps. However, by April 1913 the logbook of the A-2 showed that it was being flown almost exclusively by Lieutenant Bernard L. Smith and later with William M. McIlvain. The aircraft was redesignated E-1 on 22 September 1913. When the Marine Corps element of the Aviation Camp at Annapolis was ordered to Culebra Island, Puerto Rico, for exercises with the Advance Base Unit in January 1914, it was regarded as the Marine Section of the Navy Flying School. One of the two aircraft they operated was the E-1, while the other was the C-3 flying boat. Photos show that the C-3 was identified as a Marine Corps aircraft with the designation USMC in large letters on the tail. Both of these instances appear to only have been assignments within the Aviation Camp and not really assignment to the Marine Corps. The Marine Section, Navy Flying School was organized on 9 January 1915. There are several other early aircraft such as the H-12 and H-16 that Marines may have flown in training but that were never actually assigned to the Marine Corps.

The Marine Aeronautic Company, Advance Base Force, was organized at Marine Barracks, Philadelphia Navy Yard, Philadelphia, Pa. on 27 April 1917, with Captain Alfred A. Cunningham in command. This is the first aviation unit formed in the Marine Corps and the earliest that aircraft were officially assigned to a Marine Corps organization. On 14 October this organization was divided into two units, The First Aviation Squadron which was to train in land-based planes and the First Marine Aeronautic Company which was to train in seaplanes.

The chronological list of aircraft used by Marine aviation for the period 1913 – 2000 lists the designation, type, year assigned, manufacturer, common name, engine type, horsepower, and dimensions. The omission of a year in the master chronology indicates that no new types were received within that year. No attempt has been made to show inclusive dates within which a particular aircraft model was used nor the date a type was phased out of the Marine Corps. The numerical listing of column 1 in no way indicates the actual sequence, within a given year, that an aircraft model was assigned to the Marine Corps.

Systems for designating U.S. naval aircraft

From 1911 to 1920 there were two different systems for designating naval aircraft. The first was used from 1911 to 1914 and used a single letter to indicate the general type and manufacturer, followed by a number to indicate the individual plane of that type and manufacturer. Under this system a Curtiss amphibian was identified by the letter E while a Curtiss flying boat was designated by the letter C. The first aircraft operated by Marines were designated as the E-1 and C3.

In 1914, the system changed so that the letter A indicated a heavier-than-air craft. Within this class B stood for hydroaeroplanes and H identified flying boats. In all there were seven categories. The E-1 now became the AX-1 while the C-3 became the AB-3.

As more aircraft were obtained, the Navy found that the system had become very cumbersome and some of the later aircraft were only known by the manufacturers designation such as the JN-4 and R-6, which were from the Curtiss J and R series of aircraft. The MBT/MT (Martin Bomber Torpedo/Martin Torpedo) designations were assigned by the Navy to an aircraft procured by the Army. The DH-4, C-7 and D-7 were original foreign designations assigned by the English DeHavilland and Dutch Fokker companies.

In 1920, the Navy established a system of aircraft designations which lasted, with few modifications, till June 1962. There were four elements in this designation system as follows:

1. Aircraft Type/Class letter
2. Manufacturers Type Sequence number
3. Manufacturers Identification number
4. Modification number

The Type/Class and Manufacturers Identification letters were separated by the Manufacturers Type Sequence number. The number 1 was never used in this position. This letter number letter combination was separated from the modification number by a dash. In this case the number 1 was used for the first model accepted so that in fact the modification number was always one digit higher than the actual modification. A letter could be assigned after the modification number to show a minor change within that modification. Experimental aircraft were designated by the letter X preceding the Type/Class designation.

A typical designation is shown below.

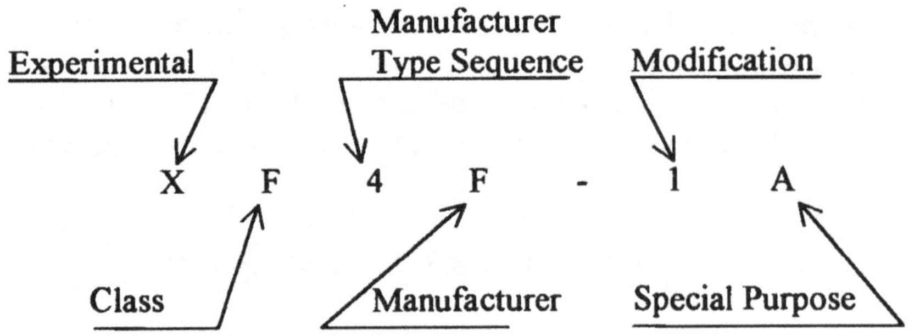

This example designates an experimental fighter, which is the fourth model built by Grumman, and is the first of that series with an engine different from the original.

In this study we are only interested in aircraft operated by the Marine Corps.

Type/Class

B	Bomber
F	Fighter
H	Hospital (1929 – 1931)
J	Transport (till 10 March 1931)
J	Utility (1931 – 1962)
JR	Utility Transport (1935 – 1962)
N	Trainer
O	Observation
R	Transport (1931 – 1962)
SB	Scout Bomber
T	Transport (1927 - 1930)
T	Torpedo
TB	Torpedo Bomber

Manufacturer Identification

A	Atlantic Aircraft Corp. (American Fokker)
A	Brewster
B	Beech
B	Bell
B	Bellanca
B	Boeing
C	Cessna
C	Curtiss Aeroplane & Motor Co.
DH	DeHavilland Aircraft Co. Ltd
DW	Dayton-Wright
D	Douglas
E	Bellanca
E	Piper
F	Grumman
G	Great Lakes
G	Goodyear
H	Howard
H	McDonnell
J	Berliner-Joyce
J	Ford
J	North American
K	Fairchild
K	Kaman
L	Grover Loening Inc.
M	General Motors
M	Glenn L. Martin Co.
O	Lockheed

P	Piasecki
P	Pitcairn Autogyro Co.
R	Ford
S	Sikorsky
T	New Standard
U	Lewis & Vought, Chance Vought, Vought
V	Lockheed
W	Wright Aeronautical Corp.
Y	Consolidated

Special Purpose Suffix

A	Target towing and photography
A	Nonfolding wings and no carrier provisions
A	Armament on normally unarmed aircraft
A	Arresting gear on normally noncarrier aircraft
A	Amphibious version
A	Land-based version of carrier aircraft
A	Built for the army Air force
B	Special armament version
B	British lend-lease version
D	Carrier operating version of noncarrier aircraft
C	British-American standardized version
C	Equipped with two cal..50 machineguns
C	Cannon armament
D	Drop tank configured
D	Special search radar
E	Special electronics version
F	Converted for use as a flagship
G	Air-sea rescue version
J	Target towing version
K	Target drone version
L	Searchlight version
M	Weather reconnaissance version
N	Night operating version (all weather)
P	Photography version
Q	Countermeasure version
R	Transport version
S	Antisubmarine version
T	Training version
U	Utility version
W	Special search version
Z	Administrative version (replaced F)

While this system had worked well for forty years, Congress decreed in 1962 that there should be just one system to designate U.S. military aircraft. The new DOD system was based heavily on the Air Force system. While few changes were made as far as the Air Force was concerned, it was a complete change in Naval Aviation. Aircraft models started with the numeral 1, except for those aircraft on hand which were used by both services, in which case the existing Air Force designation applied. Thus, the FJ-3 became the F-1C, while the SNB-5P became the RC-45J. It can not be stressed too much that the placement of the dash is critical to distinguish aircraft under the new system from those under the Navy system. For example the F4B-4 was a Boeing biplane fighter of the mid 30's, while the F-4B is an early version of the Phantom II of the 60's.

The DOD system consists of a Status Prefix Symbol (letter), a Basic Mission Symbol (letter), a Design Number (numeral), a Modified Mission Symbol (letter), a Series letter, and a Type Symbol (letter). A Design Number was assigned for each basic mission or type. New design numbers are assigned when an existing aircraft is redesigned to an extent that it no longer reflected the original configuration or capability. A Series letter is assigned to each series change of a specific basic design. To avoid confusion, the letters "I" and "O" are not used as series letters. The Series letter is always in consecutive order, starting with "A".

A typical designation is shown below.

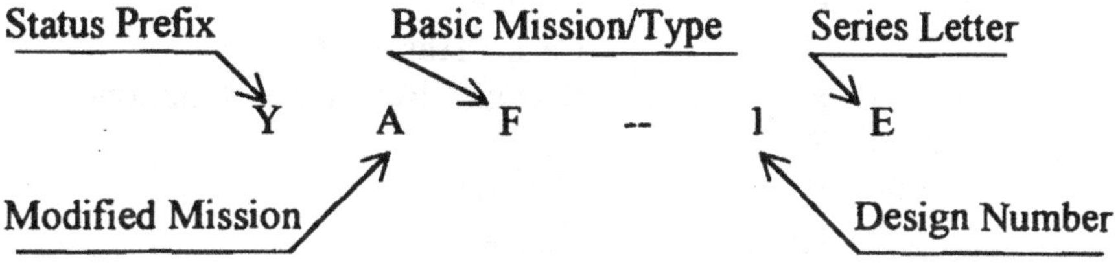

This was the Y/FJ-4B under the Navy system and the YAF-1E under the DOD system.

Status Prefix Symbols

Letter	Title
G	Permanently Grounded
J	Special Test, Temporary
N	Special test, permanent
X	Experimental
Y	Prototype
Z	Planning

Modified Mission Symbols

Letter	Title

A	Attack
C	Cargo/Transport
D	Director
E	Special Electronic Installation
H	Search/Rescue
K	Tanker
L	Cold Weather
M	Missile Carrier
Q	Drone
R	Reconnaissance
S	Antisubmarine
T	Trainer
U	Utility
V	Staff
W	Weather

Basic Mission and Type Symbols

Letter	Title
A	Attack
B	Bomber
C	Cargo/Transport
E	Special Electronic Installation
F	Fighter
H	Helicopter
K	Tanker
O	Observation
P	Patrol
S	Antisubmarine
T	Trainer
U	Utility
V	VTOL and STOL
X	Research
Z	Airship

Due to the cost of changing manuals, technical instructions, drawings etc., Navy aircraft already in the system were assigned the designation in existence under the Air Force system but also carried the old designation of the fuselage. New aircraft introduced after the change began their designations with the numeral 1. In this study I have listed the aircraft of this transitional period by their original Navy designation with the new designation beneath.

A Navy Department press release on 1 October 1941 began the official assignment of names to naval aircraft. Prior to this manufacturers did, in some cases assign a name to a specific model for publicity purposes. Perhaps the earliest was the name "Helldiver" which Curtiss assigned to their new dive bomber, the F8C/O2C series of aircraft. The name "Helldiver" was included in the October 1941 Navy press release for the SB2C-5. Vought chose the name Corsair for their carrier-based aircraft series starting with the O2U. In October 1941, the Navy assigned this name to the new Vought fighter, the F4U-1.

To fully understand the following chronological listing of aircraft it is necessary to be familiar with several of the terms used. The term "convertible" has nothing to do with a soft top automobile. Rather this means that the standard landing gear may be removed and be replaces with a single float for operation aboard battleships or cruisers. In some cases twin floats or pontoons were used.

Seaplane is the generic term under which various types of water based aircraft such as flying boats and float planes are identified. The flying boat is basically an airplane with a boat hull fuselage. Depending on the design, it could have outrigger floats under each wing for stability on the water. The float plane type has a main float attached to the fuselage by struts and outrigger floats under each wing. Some seaplanes had two floats and no outriggers. The Grumman "Duck" series married the two concepts with the main float merged with the fuselage which made it a boat hulled seaplane.

Engines used in Marine Corps aircraft range from the Curtiss inline 100 hp used in the early period to the high thrust jets with afterburner. The letter preceding the engine describes the engine type: R – radial and J – jet. Most of the other engines listed in the early period are liquid cooled inline or V configuration. The radials, which are predominate through the period prior to the end of WW II, ran from the 5 cylinder Kinner to the 28 cylinder Pratt & Whitney R-4360s. The number after the engine letter refers to the cubic displacement of the cylinders and not to horsepower i.e. R-2800. The number and letter code after cubic displacement usually indicates a modification such as fins, water injection, or other refinement i.e. R-2800-32W. In the case of jet engines, the number is assigned by sequence and does not indicate thrust or power, it serves only as identification for that specific engine.

Through the life of an aircraft numerous modifications can take place. Among these are the installation of a different engine. In this study the engine and aircraft dimensions as shown in the Navy specifications for a given aircraft have been used. However, the letter designation for Type/Class, under the Navy system, did have several changes to denote the same function. For example, the letter T was used from 1927 – 1930 to designate a transport. At that time, it was replaced by the letter R. Thus, the Fokker trimotor transport went from being a TA-1 to a RA-1, but was still the same aircraft. The designation of these aircraft was further confused by later engine changes. The TA-2 (RA-2) had a longer wing than the previous aircraft. When these RA-2s were refitted with the J-6 engines they were redesignated RA-3. The engines on the RA-1s were then changed to the 300 hp Wright J-6-9 and the aircraft were also redesignated RA-3. A TA-3 (RA-3) was purchased as such. This gave the Navy several RA-3s but with different wings. In other cases the Navy changed the type/Class designation of an aircraft as was the case of the F8C-5 being changed from a fighter to O2C-1 observation aircraft. These changes are shown in the chronology in parenthesis under the original designation. This practice is also used to show Navy aircraft that were in the system when the DOD system was adopted.

The Navy Department issued Aviation Circular Letter No. 86-45 on 17 July 1945, which changed the designation of staff transports. Prior to that time the designation "flagship", "flag aircraft", and "flag plane" had been used to designate aircraft configured primarily for special administrative purposes. These aircraft were now to be known as "administrative" aircraft. In line with this, the letter suffix "F" was replaced with the letter "Z".

Deletions and/or Corrected Designations
(Numbers as shown in original publication)

1.	AX-1	Later designation of E-1
3.	H-12	Miami training – not MarCor
4.	H-16	Miami training – not MarCor
5.	HS-2	No HS-2s built
7.	Kirkham Fighter	Not MarCor
28.	T3M-1	Not manufactured till 1925
32.	JN-6H-B	No such aircraft
56.	OC-1	Redesignation of #52 XF8C-1, f8C-1
57.	OC-2	Redesignation of #53 F8C-3
68.	RR-2	Redesignation of #64 JR-2
74.	O2C-1	Redesignation of #67 F8C-5
75.	O3U-2	Converted to SU-1 prior to MarCor issue
78.	RA-3	Redesignation of #70 TA-3
97.	SU-1	Already accounted for at #75 O3U-2
102.	SOC-1	Not MarCor
103.	F2A-1	Not built at this time
134.	FF-1	All converted to FF-2 prior to MarCor issue
140.	TBD-1	Not MarCor
170.	SNJ-1	Not MarCor
180.	GK-1	Not MarCor
181	HE-1	Early designation of #173 AE already shown
182	J4F-1	All to Coast Guard
184	JK-1	Incorrectly identified – should be J2K-1
190.	PBJ-1	No such aircraft designation
192.	PBY-1	Not MarCor
197.	R5O-1	Not MarCor
231.	PBO-1	Not MarCor
232.	R3O-2	Not MarCor
233.	R4D-2	Not MarCor
258.	PB4Y-1F	Early designation of #259 PB4Y-1Z
266.	TBM-3W	Not MarCor
267.	FT-1	Incorrectly identified – Should be F2T-1
268.	F4U-1B	Later designation of #211 F4U-1C
275.	R4D-6C	No such aircraft
276.	R4D-7	Not MarCor
278.	R5D-2Z	Redesignation of R5D-2F
300.	F7F-2P	Not MarCor
304.	R5D-4	Not MarCor
306.	F4U-4D	No –4D model built
350.	HO58-1	Incorrectly identified – Should be HO5S-1
396.	H-13D	None purchased by Navy
405.	U1A	Incorrectly identified – should be U-1A
413.	UH-34D	Incorrectly identified – Should be VH-34D

Additions

1.	E-1
2.	C-3
3.	Farman
4.	JN-4
6.	S-4B
9.	N-9H
12.	JN-4H
13.	JN-4HG
14.	JN-6H
44.	NY-1B
57.	NY-2
63.	FB-3
66.	N2C-1
67.	OL-8A
72.	XF6C-6
74.	XF7C-1
75.	O2U-2
86.	SU-1
112.	XOSU-1
129.	FF-2
138.	F2A-3P
148.	F2A-2P
253.	TBM-3P

All entries after number 463

	Designation	Type	Assigned	Manufacturer and Name	Engine Type & Horsepower	Dimensions Length	Span
1	E-1	Amphibian, 1 pusher engine, 2 place side-by-side, biplane.	1914	Curtiss	Curtiss 6 cylinder, V-8, 60 hp	27'2"	37'1"
2.	C-3	Flying-boat, 1 pusher engine, 2 place side-by-side, biplane.	1914	Curtiss	Curtiss O 75 hp	27'4"	41'8"
3.	Farman	Tractor, seaplane	1917	Farman			
4.	JN-4	Trainer, 1 engine, 2 place in tandem, biplane	1917	Curtiss "Jenny"	Curtiss OX-5 100 hp	27'4"	43'3"
5.	JN-4B	Trainer, 1 engine, 2 crew, biplane.	1917	Curtiss "Jenny"	Curtiss OX-5 100 hp	27'4"	43'3"
6.	S-4B	Scout, 1 engine, single place, biplane.	1918	Thomas-Morse	Gnome rotary 100 hp	19'10"	26'6"
7.	HS-2L	Patrol, 1 engine, 2 crew, biplane	1918	Curtiss; Low, Willard, Fowler; Naval Aircraft Factory; Standard; and others.	Liberty 12 cyl. 360 hp	39'	74'
8.	N-9	Trainer, 1 engine, 2 place in tandem, biplane, seaplane.	1918	Curtiss.	Curtiss OXX-6 100 hp	30'10"	53'4"
9.	N-9H	Trainer, 1 engine, 2 place in tandem, biplane, seaplane.	1918	Curtiss	Wright-Hispano A 150 hp	30'10"	53'4"

	Designation	Type	Assigned	Manufacturer and Name	Engine Type & Horsepower	Dimensions Length	Span
10.	R-6	Trainer, 1 engine, 2 place in tandem, biplane, seaplane.	1918	Curtiss	Curtiss V-XX 200 hp	33'5"	57'1"
11.	DH-4	Observation, 1 engine, 2 place in tandem, biplane.	1918	Dayton-Wright	Liberty 12 cyl. 360 hp	30'2"	42'6"
12.	JN-4H	Trainer, 1 engine, 2 place in tandem, biplane.	1918	Curtiss "Jenny"	Wright-Hispano 150 hp	27'	43'7'
13.	JN-4HG	Trainer, 1 engine, 2 place in tandem, biplane, for gunnery.	1918	Curtiss "Jenny"	Wright-Hispano 150 hp	27'	43'7"
14.	JN-6H	Trainer, 1 engine, 2 place in tandem, biplane, ailerons on upper and lower wings	1919	Curtiss "Jenny"	Wright-Hispano 150 hp	27'	43'7"
15.	E-1 "M" Defense	Fighter, 1 engine, single place, biplane.	1919	Standard	LeRhone rotary 80 hp	18'11"	24'
16.	DH-4B	Observation, 1 engine, 2 place in tandem, biplane.	1920	Dayton-Wright and others	Liberty 12 cyl. 400 hp	30'2"	42'
17.	HS-1	Patrol, 1 engine, 2 place, biplane, flying boat.	1920	Curtiss	Curtiss, V-XX 200 hp	39"	62'
18.	JN-6HG-I	Trainer, 1 engine, 2 place in tandem, biplane, for gunnery.	1919	Curtiss "Jenny"	Wright-Hispano Model I 150	27'4"	43' 7"

19.	VE-7	Trainer, 1 engine, 2 place in tandem, biplane.	1920	Vought and Naval Aircraft Factory	Wright Hispano E-2 180 hp	24'5"	34'1"
20.	D-7	Fighter, 1 engine, single place, biplane.	1920	Fokker	Mercedes 180 hp or BMW 185 hp	23"	27'6"
21.	C-1	Fighter, 1 engine, 2 place, biplane.	1921	Fokker	BMW 185 hp	23'8"	34'10"
22.	VE-7G	Trainer, 1 engine, single crew, biplane, for gunnery.	1921	Vought & Naval Aircraft Factory	Wright Hispano E-2 180 hp	24'5"	34'1"
23.	VE-7SF	Fighter trainer, 1 engine, single crew, biplane.	1921	Vought & Naval Aircraft Factory	Wright Hispano E-2 180 hp	24'5"	34'1"
24.	DH-4B-1	Observation, 1 engine, 2 place in tandem. Increased fuel capacity from 96 to 118 gallons, larger wheels.	1921	Dayton Wright	Liberty 12 cyl. 400 hp	30'2"	42'6"
25.	MBT	Torpedo bomber, 2 engines, 3 crew, biplane.	1921	Martin	2 Liberty 12 cyl. 400 hp	46'4"	71'5"
26.	MT	Torpedo plane, 2 engines, 3 crew, biplane.	1922	Martin	2 Liberty 12 cyl. 400 hp	45'8"	71'5"
27.	F-5L	Patrol bomber, 2 engine, 5 crew, biplane, flying boat.	1922	Curtiss, Naval Aircraft Factory and others	2 Liberty 12 cyl. 400 hp	49'4"	103'9"
28.	MB-3	Fighter, 1 engine, single crew, biplane.	1922	Thomas-Morse	Hispano 300 hp	20'	26'

	Designation	Type	Assigned	Manufacturer and Name	Engine Type & Horsepower	Dimensions Length	Span
29.	DT-2	Torpedo bomber, 1 engine, 2 place in tandem, convertible, biplane.	1923	Douglas; Naval Aircraft Factory; Lowe, Willard, Fowler.	Liberty 12 cyl. 400 hp	37'8"	50'
30.	VE-9	Observation, 1 engine, 2 place in tandem, biplane.	1923	Vought	Wright E-3 180 hp	24'6"	34'1"
31.	DH-4B-2	Observation, 1 engine, 2 place in tandem, biplane, all armament removed.	1925	Dayton Wright	Liberty 12 cyl. 400 hp	30'2"	42'5"
32.	O2B-1	Observation, 1 engine, 2 place in tandem, biplane.	1925	Boeing	Liberty 12 cyl. 400 hp	30'2"	42'6"
33.	TW-3	Trainer, 1 engine, 2 place side by side, biplane.	1925	Dayton Wright "Chummy"	Wright 180	25'11"	34'10"
34.	DT-4	Torpedo, 1 engine, single place, biplane, modified for racing.	1925	Douglas, modified by Naval Aircraft Factory	Wright T-2 525 hp	34'5"	50'
35.	VE-7H	Trainer, 1 engine, 2 place in tandem, single float seaplane, biplane.	1925	Vought	Wright E-2 180 hp	24'5"	34'2"
36.	OL-2	Observation, 1 engine, 2 place in tandem, amphibian, biplane.	1926	Loening	Liberty 12 cyl. 400 hp	33'10"	45'

37.	[illegible]-3	Fighter, 1 engine, single place, biplane, convertible.	1926	Curtiss "Hawk"	Curtiss D-12 400 hp	22'8"	31'6"
38.	FB-1	Fighter, 1 engine, single place, biplane.	1926	Boeing	Curtiss D-12 400 hp	23'6"	32'
39.	NB-1	Trainer, 1 engine, 2 place in tandem,	1926	Boeing	Wright J-4 200 hp	28'9"	36'10"
40.	NB-2	Trainer, 1 engine, 2 place in tandem, single float convertible, biplane.	1926	Boeing	Wright E-4 180 hp	28'9"	36'10"
41.	NY-1	Trainer, 1 engine, 2 place in tandem, single float convertible, biplane.	1926	Consolidated	Wright J-5 200 hp	27'11"	34'6"
42.	OD-1	Observation, 1 engine, 2 place in tandem, biplane.	1926	Douglas	Packard 1A-1500 500 hp	28'8"	39'8"
43.	XS-1	Scout, 1 engine, single place, 2-float seaplane, biplane.	1926	Cox-Klemin	Lawrence L-4S 60 hp	18'	18'2"
44.	NY-1B	Trainer, 1 engine, 2 place in tandem, biplane.	1926	Consolidated	Wright R-790 200	27'11'	40"
45.	F6C-1	Fighter, 1 engine, single place, convertible, biplane.	1927	Curtiss "Hawk"	Curtiss D-12 400 hp	22'7"	31'6"

	Designation	Type	Assigned	Manufacturer and Name	Engine Type & Horsepower	Dimensions Length	Span
46.	F6C-4	Fighter, 1 engine, single place, biplane.	1927	Curtiss "Hawk"	Pratt & Whitney R-1340 410 hp	22'6"	31'6"
47.	O2U-1	Observation, 1 engine, 2 place in tandem, convertible, biplane.	1927	Chance Vought	Pratt & Whitney R-1340-88 450 hp	24'6"	34'6"
48.	OL-4	Observation, 1 engine, 3 place in tandem, amphibian, biplane	1927	Loening	Liberty 12 cyl. 400 hp	35'1"	45'
49.	OL-6	Observation, 1 engine, 3 place in tandem, amphibian, biplane.	1927	Loening	Packard 2A-1500 475 hp	35'4"	45'
50.	TA-1 (RA-1) (RA-3)	Transport, 3 engines, 2 crew, high wing monoplane	1927	Atlantic Aircraft; (Fokker)	3 Wright J-5 220 hp 3 Wright J-6-9 300 hp	49'1"	63'4"
51.	XF6C-5	Experimental Fighter, 1 engine, single place, biplane.	1927	Curtiss "Hawk"	Pratt & Whitney R-1700 525 hp	25'5"	31'6"
52.	F7C-1	Fighter, 1 engine, single place, biplane.	1928	Curtiss "Sea Hawk"	Pratt & Whitney R-1340-B 450 hp	22'2"	32'8"

53.	XF8C-1 F8C-1 (OC-1)	Fighter bomber, 1 engine, 2 place tandem, biplane.	1928	Curtiss "Falcon"	Pratt & Whitney R-1300 425 hp	25'11"	32'
54.	F8C-3 (OC-2)	Fighter bomber, 1 engine, 2 place tandem, biplane.	1928	Curtiss "Falcon"	Pratt & Whitney R-1340 432 hp	27'11"	38'
55.	OL-8	Observation, 1 engine, 2 place in tandem, amphibian, biplane.	1928	Loening	Pratt & Whitney R-1300 425 hp	34"9"	45'
56.	TA-2 (RA-2) (RA-3)	Transport, 3 engines, 2 crew, high wing monoplane.	1928	Atlantic Aircraft; (Fokker)	2 Wright R-790 220 hp 1 Pratt & Whitney R-1340B 450 hp	48'7"	72'10"
57.	NY-2	Trainer, 1 engine, 2 place in tandem, biplane.	1928	Consolidated Aircraft	Wright R-790-8 220 hp	27'11"	40'
58.	UO-1	Observation, 1 engine, 2 place in tandem, single float convertible, biplane.	1928	Chance Vought	Wright J-3 200 hp	24'6'	34'4"
59.	UO-5	Observation, 1 engine, 2 place in tandem, single float convertible, biplane.	1928	Chance Vought	Wright J-5 220 hp	28'4"	34'4"

	Designation	Type	Assigned	Manufacturer and Name	Engine Type & Horsepower	Dimensions Length	Span
60.	XOL-8	Observation, 1 engine, 3 place in tandem, biplane, amphibian.	1928	Loening	Pratt & Whitney R-1300 425 hp	34'9"	45"
61.	FB-5	Fighter, 1 engine, single place, biplane.	1929	Boeing	Packard 2A-1500 475 hp	23'2"	32'
62.	JR-2 (RR-2)	Transport, 3 engines, 2 crew, 10 passengers, high wing monoplane.	1929	Ford "Tin Goose"	3 Wright J6-9 300 hp	49'10"	74'
63.	FB-3	Fighter, 1 engine, single place, biplane.	1929	Boeing	Packard 1A-1500 510 hp	23'2"	32'
64.	OL-3	Observation, 1 engine, 3 place in tandem, biplane, amphibian.	1929	Loening	Packard 2A-2500 475 hp	35'1"	45'
65.	XHL-1	Transport, 1 engine, 2 crew, biplane, amphibian cabin, ambulance.	1929	Loening	Pratt & Whitney R-1690 525 hp	34'9"	46'10"
66.	N2C-1	Trainer, 1 engine, 2 place in tandem, biplane	1929	Curtiss "Fledgling"	Wright R-790-8 200 hp	27'5"	39'5"
67.	OL-8A	Observation, 1 engine 2 place in tandem, biplane, amphibian.	1930	Loening	Pratt & Whitney R-1340-4 450 hp	34'9"	45'

68.	F8C-5 (O2C-1)	Fighter, 1 engine 2 place in tandem, biplane.	1930	Curtiss "Helldiver"	Pratt & Whitney R-1690 525 hp	34'9"	46'10"
69.	O2U-4	Observation, 1 engine, 2 place in tandem, biplane.	1930	Chance Vought "Corsair"	Pratt & Whitney R-1340C 450 hp	30'	36'
70.	TA-3 (RA-3)	Transport, 3 engines, 2 crew, high wing monoplane.	1930	Atlantic Aircraft Corporation	3 Wright R-975 300 hp	48'1"	63'4"
71.	XN2B-1	Trainer, 2 place in tandem, biplane.	1930	Boeing	Wright R-540 165 hp	25'8"	35'
72.	XF6C-6	Fighter, single place, monoplane, converted for racing	1930	Curtiss	Curtiss Conqueror V-1570 770 hp	Undetermined	
73.	XOC-3	Observation, 2 place in tandem, biplane.	1930	Curtiss "Falcon"	Curtiss Chieftain 600 hp	27'11	38'
74.	XF7C-1	Fighter, 1 engine, single place, biplane	1931	Curtiss "Sea Hawk"	Pratt & Whitney R-1340-F 450 hp	22'2"	32'8"
75.	O2U-2	Observation, 2 place in tandem, biplane.	1931	Vought "Corsair"	Pratt & Whitney R-1340-88	24'6"	34'6"

	Designation	Type	Assigned	Manufacturer and Name	Engine Type & Horsepower	Dimensions Length	Span
76.	NT-1	Trainer, 1 engine, 2 place in tandem, biplane.	1931	New Standard	Kinner K-5 115 hp	24'7"	30'
77.	OL-9	Observation, 1 engine, 2 place in tandem, biplane, amphibian.	1931	Loening	Pratt & Whitney R-1340C 450 hp	34'9"	45'
78.	XOP-1	Observation, 1 engine, 2 place in tandem, autogyro.	1931	Pitcarin	Wright R-975 300 hp	23'1"	Rotor 30'3"
79.	RC-1	Transport, 2 engine, 2 crew, high wing, monoplane.	1931	Curtiss "Kingbird"	2 Wright R-975 300 hp	34'10"	54'6"
80.	JR-3 (RR-3)	Transport, 3 engine, 2 crew, 10 passengers, high wing monoplane.	1931	Ford "Tin Goose"	3 Wright J6-9 300 hp	50'3"	77'10"
81.	RS-1	Transport, 2 engine, 2 crew, 7 passenger, parasol wing, amphibian.	1931	Sikorsky	2 Pratt & Whitney R-1860 575 hp	45'2"	79'9"
82.	RS-3	Transport, 2 engine, 2 crew, 8 passenger, parasol wing, amphibian.	1931	Sikorsky	2 Pratt & Whitney R-1340-C 450 hp	40'3"	71'8"
83.	T4M-1	Torpedo bomber, 1 engine, 3 place, biplane.	1931	Martin	Pratt & Whitney R-1690 525 hp	37'8"	53'

84.	RR-5	Transport, 3 engines, 2 crew, 10 passenger, high-wing, monoplane,	1931	Ford "Tin Goose"	3 Pratt & Whitney R-1340-96 450 hp	50'3"	77'10"
85.	F4B-4	Fighter, 1 engine, single place, biplane.	1932	Boeing	Pratt & Whitney R-1340-16 550 hp	20'4"	30'
86.	SU-1	Scout, 1 engine, 2 place in tandem, biplane.	1932	Chance Vought "Corsair"	Pratt & Whitney R-1690-C 600 hp	26'	36'
87.	SU-2	Scout, 1 engine, 2 place in tandem, biplane.	1932	Chance Vought "Corsair"	Pratt & Whitney R-1690-C 600 hp	26'	36'
88.	F3B-1	Fighter, 1 engine, single place, biplane.	1933	Boeing	Pratt & Whitney R-1340-80 425 hp	24'10"	33'
89.	N2C-2	Trainer, 1 engine, 2 place in tandem, biplane.	1933	Curtiss "Fledgling"	Wright R-760-90 240 hp	27'9"	39'1"
90.	RE-3 (XRE-3)	Transport, 1 engine, 2 crew, 4 passenger, high-wing, monoplane.	1933	Bellanca "Pacemaker"	Pratt & Whitney R-1349-CD 450 hp	27'10"	46'4"
91.	SU-3	Scout, 1 engine, 2 place in tandem, biplane.	1933	Chance Vought "Corsair"	Pratt & Whitney R-1690-C 600 hp	26'	36'

	Designation	Type	Assigned	Manufacturer and Name	Engine Type & Horsepower	Dimensions Length	Span
92.	F4B-3	Fighter, 1 engine, single place, biplane.	1933	Boeing	Pratt & Whitney R-1349-10 500 hp	20'	30'
93.	JF-1	Utility, 1 engine, 2 place in tandem, biplane, amphibian.	1934	Grumman "Duck"	Pratt & Whitney R-1830-62 700 hp	14'4"	39'
94.	R2D-1	Transport, 2 engine, 2 crew, 14 passenger, low-wing, monoplane.	1934	Douglas	2 Wright R-1820-12 725 hp	62'	85'
95	R4C-1	Transport, 2 engine, 2 crew, 14 passenger, biplane.	1934	Curtiss-Wright "Condor"	2 Wright R-1820-12 725 hp	50'3"	82'
96.	RR-4	Transport, 3 engines, 2 crew, 10 passengers, high-wing, monoplane.	1934	Ford "Tin Goose"	3 Pratt & Whitney R-1340-96 450 hp	50'3'	77'10"
97.	BG-1	Bomber, 1 engine, 2 place in tandem, biplane.	1935	Great Lakes	Pratt &Whitney R-1535-66 700 hp	28'9"	35'
98.	JF-2	Utility, 1 engine, 2 place in tandem, biplane, amphibian.	1935	Grumman "Duck"	Pratt & Whitney R-1820-62 700 hp	14'4'	39'

99.	O3U-6	Observation, 1 engine, 2 place in tandem, biplane.	1935	Chance Vought "Corsair"	Pratt & Whitney R-1340-12 550 hp	27'2"	36'
100.	RD-3	Transport, 2 engine, 2 crew, 7 passengers, high-wing, monoplane, boat hull amphibian.	1935	Douglas	2 Pratt & Whitney R-1340-96 450 hp	45'2"	60'
101.	F3F-1	Fighter, 1 engine, single place, biplane.	1936	Grumman	Pratt & Whitney R-1535-84 650 hp	23'5"	32'
102.	O3U-1	Observation, 1 engine, 2 place in tandem, biplane.	1936	Chance Vought "Corsair"	Pratt & Whitney R-1340-96 450 hp	29'11"	36'
103.	RD-2	Transport, 2 engine, 2 crew, 7 passengers, high-wing, monoplane, boat hull amphibian.	1936	Douglas	2 Pratt & Whitney R-1340-96 450 hp	45'3"	60'
104.	SU-4	Scout, 1 engine, 2 place in tandem, biplane.	1936	Chance Vought "Corsair"	Pratt & Whitney R-1690-42 600 hp	28'	36'
105.	XBG-1	Bomber, 1 engine, 2 place in tandem biplane.	1936	Great Lakes	Pratt & Whitney R-1535-66 700 hp	33'9"	36'

	Designation	Type	Assigned	Manufacturer and Name	Engine Type & Horsepower	Dimensions Length	Span
106.	F2F-1	Fighter, 1 engine, single place, biplane.	1937	Grumman	Pratt & Whitney R-1535-84 650 hp	21'2"	28'6'
107.	F3F-2	Fighter, 1 engine, single place, biplane.	1937	Grumman	Pratt & Whitney R-1535-84 650 hp	23'2"	32'
108.	J2F-1	Utility, 1 engine, 2 place in tandem, biplane, amphibian.	1937	Grumman "Duck"	Pratt & Whitney R-1820-08 750 hp	34'	39'
109.	JO-2	Transport, 2 engine, 2 crew, 6 passenger, low-wing monoplane.	1937	Lockheed	2 Pratt & Whitney R-985-48 400 hp	36'4"	49'6"
110.	SBC-3	Scout bomber, 1 engine, 2 place in tandem, biplane.	1937	Curtiss "Helldiver"	Pratt & Whitney R-1535-94 750 hp	28'	34'
111.	XB2G-1	Bomber, 1 engine, 2 place in tandem, biplane.	1937	Great Lakes	Pratt & Whitney R-1535-94 750 hp	28'10"	36'
112.	XOSU-1	Observation scout, 1 engine, 2 place in tandem, biplane.	1937	Vought "Corsair"	Pratt & Whitney R-1340-12 550 hp	27'2"	36'

113.	XF13C-3	Fighter, 1 engine, single place in cabin, high-wing monoplane.	1937	Curtiss	Wright XR-1510-12 700 hp	26'4"	35'
114.	J2F-2	Utility, 1 engine, 2 place in tandem, biplane, amphibian.	1937	Grumman "Duck"	Wright R-1820-30 750 hp	33'	39'
115.	JRS-1	Utility transport, 2 engines, 5 crew, high-wing monoplane, flying boat.	1938	Sikorsky	2 Wright R-1690-52 600 hp	51'1"	86'
116.	O3U-3	Observation, 1 engine, 2 place in tandem, biplane.	1938	Chance Vought "Corsair"	Pratt & Whitney R-1340-12 600 hp	31'	36'
117.	SB2U-1	Scout Bomber, 1 engine, 2 place in tandem, low-wing monoplane.	1938	Vought-Sikorsky "Vindicator"	Pratt & Whitney R-1535-80 700 hp	34'	42'
118.	SOC-3	Scout observation, 1 engine, 2 place in tandem, biplane.	1938	Curtiss-Wright "Seagull"	Pratt & Whitney R-1340-22 550 hp	31'1"	36'
119.	TG-1	Torpedo bomber, 1 engine, 3 place in tandem, biplane.	1939	Great Lakes	Pratt & Whitney R-1690-28 525 hp	34'8"	53'
120.	F3F-3	Fighter, 1 engine, single place, biplane.	1939	Grumman	Wright R-1820-22 950 hp	23'3"	32'

	Designation	Type	Assigned	Manufacturer and Name	Engine Type & Horsepower	Dimensions Length	Span
121.	J2F-2A	Utility, 1 engine, 2 place in tandem, biplane, amphibian.	1939	Grumman "Duck"	Wright R-1820-50 750 hp	34'	39'
122.	J2F-4	Utility, 1 engine, 2 place in tandem, biplane, amphibian.	1939	Grumman "Duck"	Wright R-1820-30 750 hp	34'	39'
123.	JRF-1A	Utility, 2 engine, 4 crew, high-winged monoplane, amphibian, boat hull.	1939	Grumman "Goose"	2 Pratt & Whitney R-985-48 400 hp	38'4"	49'
124.	SBC-4	Scout bomber, 1 engine, 2 place in tandem, biplane.	1939	Curtiss "Helldiver"	Wright R-1820-34 950 hp	27'5"	34'
125.	XSBC-4	Scout bomber, 1 engine, 2 place in tandem, biplane.	1939	Curtiss "Helldiver"	Wright R-1820-22	27'5"	34'
126.	R3D-2	Transport, 2 engines, 4 crew, high-wing monoplane.	1940	Douglas	2 Wright R-1820-44 1000 hp	62'2"	78'
127.	SBD-1	Scout bomber, 1 engine, 2 place in tandem, low-wing monoplane.	1940	Douglas "Dauntless"	Wright R-1820-32 1000 hp	32'	41'
128.	SNJ-2	Scout trainer, 1 engine, 2 place in tandem, low-wing monoplane.	1940	North American 'Texan"	Pratt & Whitney R-1340-56 600 hp	28'11"	42'

129.	FF-2	Fighter, 1 engine, single, biplane.	1940	Grumman "FiFi"	Wright R-1820-78 700 hp	24'6"	34'6"
130	F2A-3	Fighter, 1 engine, single place, low-wing, monoplane.	1941	Brewster "Buffalo"	Pratt & Whitney R-1820-40 1,200 hp	26'4"	35'
131.	F4F-3	Fighter, 1 engine, single place, mid-wing monoplane.	1941	Grumman "Wildcat"	Pratt & Whitney R-1830-76 1,200 hp	28'9"	38"
132.	F4F-3A	Fighter, 1 engine, single place, mid-wing monoplane.	1941	Grumman "Wildcat"	Pratt & Whitney R-1830-90 1,200 hp	28'9"	38"
133.	JRB-2	Utility transport, 2 engines, 2 crew, 4 passengers, low-wing monoplane.	1941	Beech "Voyager"	2 Pratt & Whitney R-985-50, or AN-4 400 hp	43'3"	47'8"
134.	JRF-4	Utility transport, 2 crew, 4 passenger, high-wing monoplane, amphibian, boat hull.	1941	Grumman "Goose"	2 Pratt & Whitney R-985-40 450 hp	38'4"	49'
135.	SB2U-3	Scout bomber, 1 engine, 2 place in tandem, low-wing monoplane.	1941	Vought-Sikorsky "Vindicator"	Pratt & Whitney R-1535-02 825 hp	34'4"	42'
136.	SBD-2	Scout bomber, 1 engine, 2 place in tandem, low-wing monoplane.	1942	Douglas "Dauntless"	Wright R-1820-32 1,000 hp	32'	41'

	Designation	Type	Assigned	Manufacturer and Name	Engine Type & Horsepower	Dimensions Length	Span
137.	SNJ-3	Scout trainer, 1 engine, 2 place in tandem, low-wing monoplane.	1942	North American "Texan"	Pratt & Whitney R-1340-AN-1 550 hp	29'	42'
138.	F2A-3P	Fighter, 1 engine, single place, low-wing monoplane.	1942	Brewster "Buffalo"	Pratt & Whitney R-1820-40 1,200 hp	26'4"	35'
139.	BD-2	Bomber, 2 engine, 3 crew, mid-wing monoplane.	1942	Douglas "Havoc"	2 Wright R-2600-11 1,600 hp	47'3"	61'4"
140.	F2A-2	Fighter, 1 engine, single place, low-wing monoplane	1942	Brewster "Buffalo"	Wright R-1820-40 1,200 hp	25'7"	35'
141.	F4F-3P	Fighter, 1 engine, single place, low-wing monoplane.	1942	Grumman "Wildcat"	Pratt & Whitney R-1830-76 1,200 hp	28'9"	38'
142.	LNE-1	Trainer, 2 place, shoulder-wing, monoplane.	1942	Pratt-Read	Glider		54'6"
143.	LNS-1	Trainer, 2 place, shoulder-wing, monoplane.	1942	Schweizer	Glider	23'3"	52'
144.	F4F-4	Fighter, 1 engine, single place, low-wing, monoplane.	1942	Grumman "Wildcat"	Pratt & Whitney R-1830-86 1,200 hp	28'9"	38'

145.	F4F-7	Fighter, 1 engine, single place, low-wing monoplane.	1942	Grumman "Wildcat"	Pratt & Whitney R-1830-76 1,200 hp	28'9"	38'
146.	F4U-1	Fighter, 1 engine, single place, inverted gull wing, monoplane.	1942	Vought Sikorsky "Corsair"	Pratt & Whitney R-2800-8 2,000 hp	31'11"	41'11"
147.	GB-2	Transport, 1 engine, 2 crew, 3 passengers, biplane.	1942	Beech "Traveler"	Pratt & Whitney R-985-48 400 hp	26'	32"
148.	F2A-2P	Fighter, 1 engine, single place, low-wing monoplane.	1942	Brewster "Buffalo"	Wright R-1820-40 1,200 hp	25'7"	35'
149.	GH-1	Transport, 1 engine, 1 crew, 3 passengers, high-wing monoplane.	1942	Howard "Nightingale"	Pratt & Whitney R-985-AN-6 400 hp	25'8"	38'
150	J2F-3	Utility, 1 engine, 2 place in tandem, biplane, amphibian boat hull.	1942	Grumman "Duck"	Wright R-1820-30 750 hp	34'	39'
151.	J2F-5	Utility, 1 engine, 2 place in tandem, biplane, amphibian boat hull.	1942	Grumman "Duck"	Wright R-1820-50 850 hp	45'	39'
152.	JE-1	Utility transport, 1 engine, 2 crew, 3 passengers, high-wing monoplane.	1942	Bellanca "Pacemaker"	Pratt & Whitney R-1340-27 570 hp	28'6"	50'6"

	Designation	Type	Assigned	Manufacturer and Name	Engine Type & Horsepower	Dimensions Length	Span
153.	JRF-1	Utility transport, 2 engines, 2 crew, 5 passengers, high-wing amphibian.	1942	Grumman "Goose"	2 Pratt & Whitney R-985-AN-6 450 hp	38'4"	49'
154.	JRF-5	Utility transport, 2 engines, 2 crew, 5 passengers, high-wing amphibian.	1942	Grumman "Goose"	2 Pratt & Whitney R-985-AN-6 450 hp	38'4"	49'
155.	N2T-1	Trainer, 1 engine, 2 place in tandem, low-wing monoplane.	1942	Timm "Tutor"	Continental R-670-4 220 hp	24'10"	36'
156.	N3N-3	Trainer, 1 engine, 2 place in tandem, biplane.	1942	Naval Aircraft Factory "Yellow Peril"	Wright R-760-2 235 hp	28'4"	34'
157.	NE-1	Trainer, 1 engine, 2 place in cabin, high-wing monoplane.	1942	Piper "Grasshopper"	Continental O-170 65 hp	22'3"	5'3"
158.	OS2N-1	Observation scout, 1 engine, 2 place in tandem, low-wing monoplane, convertible, single float.	1942	Naval Aircraft Factory "Kingfisher"	Pratt & Whitney R-985-48 450 hp	33'8"	35'11"
159.	OS2U-2	Observation scout, 1 engine, 2 place in tandem, low-wing monoplane, convertible, single float.	1942	Vought-Sikorsky "Kingfisher"	Pratt & Whitney R-985-50 450 hp	33'8"	35'11"

160.	OS2U-3	Observation scout, 1 engine, 2 place in tandem, low-wing monoplane, convertible, single float.	1942	Vought-Sikorsky "Kingfisher"	Pratt & Whitney R-985-AN-2 450 hp	33'8"	35'11"
161.	PB4Y-1	Patrol bomber, 4 engines, 6-10 crew, high-wing monoplane.	1942	Consolidated "Liberator"	4 Pratt & Whitney R-1830-43 1,200 hp	64'4"	110'
192.	PBY-5A	Patrol bomber, 2 engines, 7 crew, parasol-wing amphibian, boat hull.	1942	Consolidated, "Catalina"	2 Pratt & Whitney R-1830-82 1,200 hp	65' 1"	104'
163.	R4D-1	Transport, 2 engines, 3 crew, cargo, low-wing monoplane.	1942	Douglas "Skytrain"	2 Pratt & Whitney R-1830-92 1,200 hp	64'6"	95'
164.	R4D-3	Transport, 2 engines, 3 crew, 20 passenger, low-wing monoplane.	1942	Douglas "Skytrooper"	2 Pratt & Whitney R-1830-92 1,200 hp	64'6"	95'
165.	R4D-4	Transport, 2 engines, 3 crew, 20 passenger, low-wing monoplane.	1942	Douglas "Skytrooper"	2 Pratt & Whitney R-1830-92 1,200 hp	64'6"	95'
166.	R5O-4	Transport, 2 engines, 4 crew, 4 passengers, low-wing monoplane.	1942	Lockheed "Lodestar"	2 Wright R-1820 1,200 hp	49'10"	65'6"
167.	SB2A-4	Scout bomber, 1 engine, 2 crew, low-wing monoplane.	1942	Brewster "Buccaneer"	Wright R-2600-8 1,700 hp	39'2"	47'

	Designation	Type	Assigned	Manufacturer and Name	Engine Type & Horsepower	Dimensions	
						Length	Span
168.	SBD-3	Scout bomber, 1 engine, 2 crew, low-wing monoplane.	1942	Douglas "Dauntless"	Wright R-1820-52 1,000 hp	33'	41'6"
169.	SBD-4	Scout bomber, 1 engine, 2 crew, low-wing monoplane.	1942	Douglas "Dauntless"	Wright R-1820-52 1,000 hp	33'	41'6"
170.	SNC-1	Scout trainer, 1 engine, 2 crew, low-wing monoplane.	1942	Curtiss "Falcon"	Wright R-974 420 hp	28'6"	35'
171.	SNJ-4	Scout trainer, 1 engine, 2 crew, low-wing monoplane.	1942	North American "Texan"	Pratt & Whitney R-1340-AN-1 550 hp	29'	42'
172.	TBF-1	Torpedo bomber, 1 engine, 3 crew, mid-wing monoplane.	1942	Grumman "Avenger"	Wright R-2600-8 1,700 hp	41'	54'2"
173.	F4F-7P	Fighter, 1 engine, 1 crew, low-wing monoplane.	1942	Grumman "Wildcat"	Pratt & Whitney R-1830-70 1,200 hp	28'9"	38'
174.	HE-1 (AE-1)	Ambulance utility, 1 engine, 1 crew, 1 litter, high-wing monoplane.	1942	Piper	Continental A-75-8 75 hp	22'6"	35'6"
175.	F6F-3	Fighter, 1 engine, 1 crew, low-wing monoplane.	1943	Grumman "Hellcat"	Pratt & Whitney R-2800-10 2,000 hp	33'7"	42'10"

176.	F6F-3N	Night fighter, 1 engine, 1 crew, low-wing monoplane.	1943	Grumman "Hellcat"	Pratt & Whitney R-2800-10W 2,000 hp	33'7"	42'10"
177.	R5C-1Z	Transport, 2 engine, 2 crew, 40 passenger, land, low-wing monoplane.	1943	Curtiss "Commando"	2 Pratt & Whitney R-2800-51 2,000 hp	76'4"	108'1"
178.	FG-1	Fighter, 1 engine, 1 crew, inverted gull-wing monoplane.	1943	Goodyear "Corsair"	Pratt & Whitney R-2800 2,000 hp	33'4"	41'
179.	FM-1	Fighter, 1 engine, 1 crew, mid-wing monoplane	1943	General Motors "Wildcat"	Wright R-1820-56 1,350 hp	28'11"	38'
180.	GB-1	Utility transport, 1 engine, 2 crew, 3 passenger, biplane.	1943	Beech "Traveler	Pratt & Whitney R-985-48 400 hp	26'	32'
181.	GH-2	Utility transport, 1 engine, 1 crew, 1 passenger, 2 litter, high-wing monoplane	1943	Howard "Nightingale"	Pratt & Whitney R-985-AN-6 400 hp	26'	38"
182.	J4F-2	Utility, 2 engine, 1 crew, 3 or 4 passengers, amphibian high-wing monoplane, boat hull.	1943	Grumman "Widgeon"	2 Ranger 6-440C-5 200 hp	31'	40'
183.	J2K-1	Utility, 1 engine, 2 crew, high-wing monoplane.	1943	Fairchild	Warner R-500 165 hp	23'9"	36'4"

	Designation	Type	Assigned	Manufacturer and Name	Engine Type & Horsepower	Dimensions Length	Span
184.	JM-1	Utility, 2 engine, 3 crew, high-wing monoplane.	1943	Martin "Marauder"	2 Pratt & Whitney R-2800-43 2,000 hp	58'3"	71'
185.	JRB-1	Utility transport, 2 engine, 3 crew, 4 passenger, low-wing monoplane.	1943	Beech "Expeditor"	2 Pratt & Whitney R-985-50, or 48 400 hp	34'3"	47'8"
186.	JRC-1	Utility transport, 2 engine, 2 crew, 3 passenger, low-wing monoplane.	1943	Cessna "Bobcat"	2 Jacobs R-755 450 hp	32'9"	41'11"
187.	NH-1	Trainer, 1 engine, 2 crew, high-wing monoplane.	1943	Howard "Nightingle"	Pratt & Whitney R-985-AN-6 450 hp	26'	38'
188.	NP-1	Trainer, 1 engine, 2 place in tandem, biplane.	1943	Spartan	Lycoming R-680-8 220 hp	24'2"	33'8"
189.	PBJ-1D	Patrol-bomber, 2 engine, 5 crew, high-wing monoplane.	1943	North American "Mitchell"	2 Wright R-2600-13 1600 hp	52'10"	67'6"
190.	PV-1	Patrol, 2 engine, 4 crew, mid-wing monoplane.	1943	Vega "Ventura"	2 Pratt & Whitney R-2800-31 2,000 hp	51'2"	65'6"

191.	R4D-5 (C-47H)	Transport, 2 engine, 3 crew, 28 passenger, cargo, low-wing monoplane.	1943	Douglas "Skytrain"	2 Pratt & Whitney R-1830-92 1,200 hp	63'10"	95'
192.	R5C-1	Transport, 2 engine, 2 crew, 40 passengers, cargo, double deck, low-wing monoplane.	1943	Curtiss "Commando"	2 Pratt & Whitney R-2800-51 2,000 hp	76'4"	108' 1"
193.	R5D-5 (C-54S)	Transport, 4 engine, 6 crew, 49 passenger, cargo, low-wing monoplane.	1943	Douglas "Skymaster"	4 Pratt & Whitney R-2000-9 1,350 hp	93'11"	117'6"
194.	R5O-5	Transport 2 engine, 4 crew 12 passenger, low-wing monoplane.	1943	Lockheed "Lodestar"	2 Wright R-1820 1,000 hp	49'9"	65'6"
195.	R5O-6	Transport, 2 engine, 3 crew, 18 passenger, low-wing monoplane.	1943	Lockheed "Lodestar"	2 Wright R-1820 1,000 hp	49'9"	65'6"
196.	SB2C-1	Scout-bomber, 1 engine, 2 crew, low-wing, split flat, monoplane.	1943	Curtiss "Helldiver"	Wright R-2600 1,750 hp	35' 8"	49' 9"
197.	SBD-3P	Scout-bomber, 1 engine, 2 crew, low-wing, perforated flap, monoplane.	1943	Douglas "Dauntless"	Wright R-18200-52 1,000 hp	33'	41'6"
198.	SBD-4P	Scout-bomber, 1 engine, 2 crew, low-wing, perforated flap, monoplane.	1943	Douglas "Dauntless"	Wright R-1820-52 1,000 hp	33'	41'6"

	Designation	Type	Assigned	Manufacturer and Name	Engine Type & Horsepower	Dimensions Length	Span
199.	SBD-5	Scout-bomber 1 engine, 2 crew, low-wing, perforated flap, monoplane.	1943	Douglas "Dauntless"	Wright R-1820-60 1,200 hp	33'	41'6"
200.	SBD-5A	Scout-bomber, 1 engine, 2 crew, low-wing, perforated flap, monoplane.	1943	Douglas "Dauntless"	Wright R-1820-60 1,200 hp	33'	41'6"
201.	SNB-1	Trainer, 2 engine, 4 crew, low-wing, monoplane.	1943	Beech "Kansan"	2 Pratt & Whitney R-985-AN-3 450 hp	34'3"	47'8"
202.	SNB-2	Trainer, 2 engine, 4 or 5 crew, low-wing, monoplane.	1943	Beech "Navigator"	2 Pratt & Whitney R-985-AN-4 450 hp	34'3"	47'8"
203.	SNJ-5	Trainer, 1 engine, 2 crew, low-wing, monoplane.	1943	North American "Texan"	Pratt & Whitney R-1340-AN-1 550 hp	29'	42'
204.	SNV-1	Scout-trainer, 1 engine, 2 crew, low-wing, monoplane.	1943	Vultee "Valiant"	Pratt & Whitney R-985-AN-1 450 hp	28'8"	42'
205.	TBM-1	Torpedo-bomber, 1 engine, 3 crew, mid-wing, monoplane.	1943	General Motors "Avenger"	Wright R-2600-8 1,700 hp	40'	54'2"

206.	F3A-1	Fighter, 1 engine, 1 crew, land, inverted gull-wing, monoplane.	1943	Brewster "Corsair"	Pratt & Whitney R-2800-8 2,000 hp	33'4"	41'
207.	F4U-1C	Fighter, 1 engine, 1 crew, land, inverted gull-wing, monoplane.	1943	Vought "Corsair"	Pratt & Whitney R-2800-8 2,000 hp	33'4"	41'
208.	F4U-1D	Fighter, 1 engine, 1 crew, land, inverted gull-wing, monoplane.	1943	Vought "Corsair"	Pratt & Whitney R-2800-8W 2,000 hp	33'4"	41'
209.	F6F-3E	Fighter, 1 engine, 1 crew, land, low-wing, monoplane.	1944	Grumman "Hellcat"	Pratt & Whitney R-2800-10 2,000 hp	33'6"	42'10"
210	F6F-3P	Fighter, 1 engine, 1 crew, land, low-wing, monoplane.	1944	Grumman "Hellcat"	Pratt & Whitney R-2800-10W 2,000 hp	33'6"	42'10"
211.	F6F-5	Fighter, 1 engine, 1 crew, land, low-wing, monoplane.	1944	Grumman "Hellcat"	Pratt & Whitney R-2800-10W 2,000 hp	33'6"	42'10"
212.	F6F-5N	Fighter, 1 engine, 1 crew, land, low-wing, monoplane.	1944	Grumman "Hellcat"	Pratt & Whitney R-2800-10W 2,000 hp	33'6"	42'10"
213.	F6F-5P	Fighter, 1 engine, 1 crew, land, low-wing, monoplane.	1944	Grumman "Hellcat"	Pratt & Whitney R-2800-10W 2,000 hp	33'6"	42'10"

	Designation	Type	Assigned	Manufacturer and Name	Engine Type & Horsepower	Dimensions Length	Span
214.	F7F-1N	Fighter, 2 engine, 1 crew, land, high mid-wing, monoplane.	1944	Grumman "Tigercat"	Pratt & Whitney R-28000-22W 2,100 hp	45'4"	51'6"
215.	FG-1A	Fighter, 1 engine, 1 crew, land, inverted gull-wing, monoplane. Fixed wings.	1944	Goodyear "Corsair"	Pratt & Whitney R-2800-9 2,000 hp	33'4"	41"
216.	GH-3	Transport, 1 engine, 1 crew, 3 passenger, land, high-wing, monoplane	1944	Howard "Nightingale"	Pratt & Whitney R-985 450 hp	25'8"	38'
217.	J2F-6	Utility, 1 engine, 2 crew, amphibian, biplane.	1944	Columbia "Duck"	Wright R-1820-54 900 hp	34'	39'
218.	JRB-3	Utility transport, 2 engine, 2 crew, 3 passenger, land, monoplane.	1944	Beech "Navigator"	2 Pratt & Whitney R-985-AN-3, or 4 450 hp	34'3"	47'8"
219.	JRB-4	Utility-transport 2 engine, 2 crew, 7 passenger, land, monoplane.	1944	Beech "Navigator"	2 Pratt & Whitney R-985-AN-4 450 hp	34'3'	47'8"
220.	N2S-3	Trainer, 1 engine, 2 crew, land, biplane.	1944	Stearman "Kaydet	Continental R-670-4 220 hp	25'	32'2"
221.	N2S-5	Trainer, 1 engine, 2 crew, land, biplane.	1944	Stearman "Kaydet"	Lycoming R-680-17 220 hp	25'	32'2"

222.	OY-1	Observation, 1 engine, 2 crew, high wing.	1944	Consolidated-Vultee "Sentinel"	Lycoming O-435-C 190 hp	24'	34'
223.	PBJ-1C	Patrol-bomber, 2 engine, 5 crew, land, high-wing, monoplane.	1944	North American "Mitchell"	2 Wright R-2600-13 1,450 hp	54'	67'5"
224.	PBJ-1G	Patrol-bomber, 2 engine, 5 crew, land, high-wing, monoplane, first with 75mm cannon.	1944	North American "Mitchell"	2 Wright R-2600-13 1,450 hp	54'	67'5"
225.	PBJ-1H	Patrol-bomber, 2 engine, 5 crew, land, high-wing, monoplane, 75 mm cannon.	1944	North American "Mitchell"	2 Wright R-2600-13 1,450 hp	54'	67'5"
226.	PBJ-1J	Patrol-bomber, 2 engine, 5 crew, land, high-wing, monoplane.	1944	North American "Mitchell"	2 Wright R-2600-13 1,450 hp	54'	67'5"
227.	R4D-6 (C-47J)	Transport, 2 engine, 3 crew, land, low-wing, monoplane.	1944	Douglas "Skytrooper"	2 Pratt & Whitney R-1830-90C 1,200 hp	64'5"	95'
228.	SB2A-2	Scout-bomber, 1 engine, 2 crew, land, low-wing, monoplane.	1944	Brewster "Bucaneer"	Wright R-2600-8 1,700 hp	39'2"	47'
229.	SB2C-1A	Scout-bomber, 1 engine, 2 crew, land, low-wing, split flaps, monoplane.	1944	Curtiss-Wright "Helldiver"	Wright R-2600-8 1,700 hp	36'4"	49'9"

	Designation	Type	Assigned	Manufacturer and Name	Engine Type & Horsepower	Dimensions Length	Span
230.	SB2C-1C	Scout-bomber, 1 engine, 2 crew, land, low-wing, split flaps, monoplane.	1944	Curtiss-Wright "Helldiver"	Wright R-2600-8 1,700 hp	36'4"	49'9"
231.	SB2C-3	Scout-bomber, 1 engine, 2 crew, land, low-wing, split flaps, monoplane.	1944	Curtiss-Wright "Helldiver"	2 Wright R-2600-20 1,700 hp	36'4"	49'9"
232.	SB2C-4	Scout-bomber, 1 engine, 2 crew, land, low-wing, split flaps, monoplane.	1944	Curtiss-Wright "Helldiver"	2 Wright R-2600-20 1,700 hp	36'3"	49'9"
233.	SB2C-4E	Scout-bomber, 1 engine, 2 crew, land, low-wing, monoplane.	1944	Curtiss-Wright "Helldiver"	2 Wright R-2600-20 1,700 hp	36'3"	49'9"
234.	SBD-6	Scout-bomber, 1 engine, 2 crew, land, low-wing, monoplane, perforated flaps.	1944	Douglas "Dauntless	Wright R-1820-66 1,350 hp	33'	41'7"
235.	SBF-1	Scout-bomber, 1 engine, 2 crew, land, low-wing, split flaps, monoplane.	1944	Fairchild "Helldiver"	Wright R-2600-8 1,700 hp	36'4"	49'9"
236.	SBF-3	Scout-bomber, 1 engine, 2 crew, land, low-wing, split flaps, monoplane.	1944	Fairchild "Helldiver"	Wright R-2600-20 1,700 hp	36'4"	49'9"

237.	SBW-3	Scout-bomber, 1 engine, 2 crew, land, low-wing, split flaps, monoplane.	1944	Canadian Car and Foundry Company "Helldiver"	Wright R-2600-20 1,700 hp	36'4"	49'9"
238.	SNB-2C	Trainer, 2 engine, 5 crew, low-wing, monoplane.	1944	Beech "Navigator"	2 Pratt & Whitney R-985 450 hp	34'3"	47'8"
239.	TBF-1C	Torpedo-bomber, 1 engine, 3 crew, land, mid-wing, monoplane	1944	Grumman "Avenger"	Wright R-2600-8 1,700 hp	41'	54'2"
240.	TBM-1C	Torpedo-bomber, 1 engine, 3 crew, land, mid-wing, monoplane.	1944	General Motors "Avenger"	Wright R-2600-8 1,700 hp	41'	54'2"
241.	TBM-3	Torpedo-bomber, 1 engine, 3 crew, land, mid-wing, monoplane	1944	General Motors "Avenger"	Wright R-2600-8 1,700 hp	41'	54'2"
242.	F4U-4	Fighter, 1 engine, 1 crew, land/carrier, inverted gull-wing, monoplane	1945	Vought "Corsair"	Pratt & Whitney R-2800-18W 2,000 hp	33'8"	41'
243.	F6F-5E	Fighter, 1 engine, 1 crew, land, low-wing, monoplane.	1945	Grumman "Hellcat"	Pratt & Whitney R-2800-10W 1,200 hp	33'7"	42'10"
244.	F7F-2N	Fighter, 2 engine, 2 crew, land, high mid-wing, monoplane.	1945	Grumman "Tigercat"	2 Pratt & Whitney R-2800-22W 2,200 hp	31'2"	51'6"

	Designation	Type	Assigned	Manufacturer and Name	Engine Type & Horsepower	Dimensions Length	Span
245.	F7F-2P	Fighter, 2 engine, 2 crew, land, high mid-wing, monoplane	1945	Grumman "Tigercat"	2 Pratt & Whitney R-2800-22W 2,200 hp	31'2"	51'6"
246.	F7F-3	Fighter, 2 engine, 1 crew, land, high mid-wing, monoplane.	1945	Grumman "Tigercat"	2 Pratt & Whitney R-2800-22W 2,200 hp	31'2"	51'6"
247.	F7F-3N	Fighter, 2 engine, 2 crew, land, high mid-wing, monoplane.	1945	Grumman "Tigercat"	2 Pratt & Whitney R-2800-22W 2,200 hp	31'2"	51'6"
248.	F7F-3P	Fighter, 2 engine, 1 crew, land, high mid-wing, monoplane	1945	Grumman "Tigercat"	2 Pratt & Whitney R-2800-22W 2,200 hp	31'2"	51'6"
249.	FG-1D	Fighter, 1 engine, 1 crew, land, inverted gull-wing, monoplane	1945	Goodyear "Corsair"	Pratt & Whitney R-2800-8 2,000 hp	33'5"	41'
250.	FM-2	Fighter, 1 engine, 1 crew, land, mid-wing monoplane.	1945	General Motors "Wildcat"	Pratt & Whitney R-1820-56 1200 hp	28'10"	38'
251.	PB4Y-1F (PB4Y-1Z)	Patrol-bomber, 4 engine, 6 crew, 10 passengers, land, high-wing, monoplane	1945	Consoladated-Vultee "Liberator"	4 Pratt & Whitney R-1830-43 1,200 hp	67'2"	110'

252	SBF-4E	Scout-bomber,1 engine, 2 crew, land, low-wing, monoplane, split flaps.	1945	Fairchild "Helldiver"	Wright R-2600-20 1,900 hp	36'3"	49'9"
253.	TBM-3P	Torpedo-bomber, 1 engine, 3 crew, land, mid-wing monoplane.	1945	Grumman "Avenger"	Wright R-2600-20 1,600 hp	41'	54'2"
254.	SBW-4E	Scout-bomber, 1 engine, 2 crew, land, low-wing monoplane, split flaps	1945	Canadian Car and Foundry Company "Helldiver"	Wright R-2600-20 1,600 hp	36'3"	49'9"
255.	SNB-3H	Trainer, 2 engine, 2 crew, 2 passenger, 2 litter, land, low-wing, monoplane.	1945	Beech "Navigator"	2 Pratt & Whitney R-985 450 hp	34'3"	47'8"
256.	R4D-5R	Transport, 2 engine, 3 crew, 21 passenger, land, low-wing, monoplane.	1945	Douglas "Skytrain"	2 Pratt & Whitney R-1830-92 1,200 hp	64'5"	95'
257.	SB2C-5	Scout-bomber, 1 engine, crew, land, low-wing monoplane, split flaps.	1945	Curtiss "Helldiver"	Wright R-2600-20 1,600 hp	35'8"	49'9"
258.	SNJ-6	Scout-trainer, 1 engine, 2 crew, land, low-wing, monoplane.	1945	North American "Texan"	Pratt & Whitney R-1340-AN-1 500 hp	29'	42'
259.	TBM-3E	Torpedo-bomber, 1 engine, 3 crew, land, mid-wing, monoplane	1945	General Motors "Avenger"	Wright R-2600-20 1,600 hp	41'	54'2"

	Designation	Type	Assigned	Manufacturer and Name	Engine Type & Horsepower	Dimensions Length	Span
260.	F2T-1	Fighter, 2 engine, 3 crew, land, mid-wing, monoplane, twin-boom tail.	1945	Northrop "Black Widow"	2 Pratt & Whitney R-2800-10 2,000 hp	48'11"	66'
261.	R5D-2F (R5D-2Z) (VC-54P)	Transport, 4 engine, 6 crew, 22 passenger, land, low-wing monoplane	1945	Douglas "Skymaster"	4 Pratt & Whitney R-2000-7 1,250 hp	93;11"	117'6"
262.	F7F-1	Fighter, 2 engine, 1 crew, land, high mid-wing, monoplane.	1945	Grumman "Tigercat"	2 Pratt & Whitney R-2800-22W 2,000 hp	45'4"	51'6"
263.	F4U-4P	Fighter, 1 engine, 1 crew, land, inverted gull-wing, monoplane.	1946	Vought "Corsair"	Pratt & Whitney R-2800-18W 2,000 hp	33'8"	41'
264.	F7F-4N	Fighter, 2 engine, 2 crew, land, high mid-wing, monoplane.	1946	Grumman "Tigercat"	2 Pratt & Whitney R-2800-34, 2,000 hp	45'4"	51'6"
265.	JM-2	Utility, 2 engine, 3 crew, land, high-wing, monoplane.	1946	Martin "Marauder"	2 Pratt & Whitney R-2800-43 2,000 hp	56'6"	71'
266.	JRB-5	Utility-transport, 2 engine, 2 crew, 5 passenger, land, low-wing, monoplane.	1946	Beech "Expeditor"	2 Pratt & Whitney R-985 450 hp	34'3"	47'8"

267.	PB4Y-2	Patrol-bomber, 4 engine, 11 crew, land, high-wing, monoplane.	1946	Consolidated-Vultee "Privateer"	4 Pratt & Whitney R-1830-94 1,200 hp	74'	110'
268.	R5D-3 (C-54Q)	Transport, 4 engine, 6 crew, 40-50 passenger, cargo, land, low-wing monoplane.	1946	Douglas "Skymaster"	4 Pratt & Whitney R-2000-11 1,350 hp	93'11'	117'6"
269.	R5D-3Z (VC-54Q)	Transport, 4 engine, 6 crew, 22 passenger, land, low-wing monoplane.	1946	Douglas "Skymaster"	4 Pratt & Whitney R-2000-11 1,350 hp	93'11"	117'6"
270.	RY-3	Transport, 4 engine, 4 crew, 28-36 passenger, cargo, land, high-wing monoplane.	1946	Consolidated-Vultee "Liberator-Express"	4 Pratt & Whitney R-1830-94 1,200 hp	74'7"	110'
271.	SNB-3	Trainer, 2 engine, 4 crew, land, low-wing monoplane.	1946	Beech "Navigator"	2 Pratt & Whitney R-985 450 hp	34'3"	47'8"
272.	XF4U-4	Fighter, 1 engine, 1 crew, land, inverted gull-wing, monoplane.	1946	Vought "Corsair"	Pratt & Whitney R-2800-18W 2,100 hp	33'4"	41'
273.	F4U-4B	Fighter, 1 engine, 1 crew, land, inverted gull-wing, monoplane.	1947	Vought "Corsair"	Pratt & Whitney R-2800-18 2,000 hp	33'4"	41'
274.	FH-1	Fighter (jet), 2 engine, 1 crew, land, low-wing monoplane.	1947	McDonnell "Phantom"	2 Westinghouse J-30 3,000 lbs.. thrust	38'9"	40'9"

	Designation	Type	Assigned	Manufacturer and Name	Engine Type & Horsepower	Dimensions Length	Span
275.	JD-1	Utility, 2 engine, 3 crew, land, high-wing monoplane	1947	Douglas "Invader"	2 Pratt & Whitney R-2800-71 2,000 hp	50'9"	70'
276.	JRB-6	Utility-transport, 2 engine, 2 crew, 5 passenger, land, low-wing monoplane.	1947	Beech "Expeditor"	2 Pratt & Whitney R-985 450 hp	34'3"	47'8"
277.	R4D-6R	Transport, 2 engine, 3 crew, 21 passenger, land, low-wing monoplane.	1947	Douglas "Skytrain"	2 Pratt & Whitney R-1830-92 1,200 hp	64'5'	95'
278.	R5O-3	Transport, 2 engine, 4 crew, 4 passenger, land, low-wing monoplane.	1947	Lockheed "Lodestar"	2 Pratt & Whitney R-1830	49'9"	65'6"
279.	SBW-5	Scout-bomber, 1 engine, 2 crew, land, low-wing monoplane, split flap	1947	Canadian Car and Foundry Company "Helldiver"	Wright R-2600-20 1,600 hp	36'3"	49'9"
280.	SNB-2P	Trainer, 2 engine, 5 crew, land, low-wing monoplane.	1947	Beech "Navigator"	2 Pratt & Whitney R-985 450 hp	34'3"	47'8"
281.	SNB-3E	Trainer, 2 engine, 5 crew, land, low-wing monoplane.	1947	Beech "Navigator"	2 Pratt & Whitney R-985 450 hp	34'3"	47'8"

282.	SNB-3N	Trainer, 2 engine, 5 crew, land, low-wing monoplane.	1947	Beech "Navigator"	2 Pratt & Whitney R-985 450 hp	34'3"	47'8"
283.	SNJ-4C	Trainer, 1 engine, 2 crew, land/carrier, low-wing monoplane.	1947	North American "Texan"	Pratt & Whitney R-1340- AN-1 550 hp	29'	42'
284.	F4U-4N	Fighter, 1 engine, 1 crew, land, inverted gull-wing monoplane.	1948	Vought "Corsair"	Pratt & Whitney R-2800-18W 2,000 hp	33'8"	41'
285.	F4U-5	Fighter, 1 engine, 1 crew, land, inverted gull-wing monoplane.	1948	Vought "Corsair"	Pratt & Whitney R-2800-32W 2,000 hp	33'8"	41'
286.	F4U-5P	Fighter, 1 engine, 1 crew, land, inverted gull-wing monoplane.	1948	Vought "Corsair"	Pratt & Whitney R-2800-32W 2,000 hp	33'8"	41'
287.	HO3S-1	Helicopter, observation-utility, 1 engine, 1 crew, 3 passenger, land, 3 blade rotor.	1948	Sikorsky	Pratt & Whitney R-985 450 hp	57'	Rotor 48'
288.	HRP-1	Helicopter, transport, 1 engine, 2 crew, 8 passenger, cargo, land, twin 3 blade rotors.	1948	Piasecki "Rescuer"	Pratt & Whitney R-1340 600 hp	48'	Rotor 41'

	Designation	Type	Assigned	Manufacturer and Name	Engine Type & Horsepower	Dimensions Length	Span
289.	HTL-2	Helicopter, trainer, 1 engine, 2 crew, land, 2 blade rotor.	1948	Bell	Franklin 6V4-178-B-32 178 hp	41'2"	Rotor 35'
290.	TO-1	Trainer, Jet, 1 engine, 2 crew, land, low-wing monoplane.	1948	Lockheed "Shooting Star"	Allison J-33- 4,600 lbs.. thrust	34'6"	38'10"
291.	F4U-5N	Fighter, 1 engine, land, inverted gull-wing monoplane.	1949	Vought "Corsair"	Pratt & Whitney R-2800-32W 2,000 hp	34'6"	41'
292.	F9F-2	Fighter, jet, 1 engine, 1 crew, land, low wing monoplane.	1949	Grumman "Panther"	Pratt & Whitney J-42P-2 5,750 lbs.. thrust	40'	38'
293.	OY-2	Observation, 1 engine, 2 crew high-wing monoplane.	1949	Consolidated-Vultee "Sentinel"	Lycoming O-435C 190 hp	24'	34'
294.	SNB-3Q	Trainer, 2 engine, 5 crew, land, low-wing monoplane.	1949	Beech "Navigator"	2 Pratt & Whitney R-985 450 hp	34'3"	47'8"
295.	SNB-5 (TC-45J) (UC-45J)	Trainer, 2 engine, 2 crew, 4 passenger, land, low-wing monoplane.	1949	Beech "Navigator"	2 Pratt & Whitney R-985 450 hp	34'3"	47'8"
296.	AD-3W	Attack, 1 engine, 3 crew, land, low-wing monoplane.	1950	Douglas "Skyraider"	Wright R-3350-14 2,500 hp	39'4"	50'

297.	F2H-2	Fighter, jet, 2 engine, 1 crew, land, low-wing monoplane.	1950	McDonnell "Banshee"	2 Westinghouse J34- WE-34 3,000 lbs. thrust	40'2"	41'7"
298.	F9F-2B	Fighter, jet, 1 engine, 1 crew, land, low-wing monoplane.	1950	Grumman "Panther"	Pratt & Whitney J-42-P 5,750 lbs. thrust	40'	38'
299.	HRP-2	Helicopter, transport, 1 engine, 2 crew, 8 passenger, cargo, land, twin 3 blade rotors.	1950	Piasecki "Rescuer"	Pratt & Whitney R-1340-AN-1 600 hp	48'	Rotor
300	HTL-1	Helicopter, trainer, 1 engine, 2 crew, land, 2 blade rotor.	1950	Bell	Franklin V4-178-B-32 178 hp	41' 41'2"	Rotor 35'
301.	P2V-2	Patrol, 2 engine, 7 crew, land, high-wing monoplane.	1950	Lockheed "Neptune"	2 Wright R-3350-24W 2,500 hp	76'10"	100'
302.	P2V-3	Patrol, 2 engine, 7 crew, land, high-wing, monoplane.	1950	Lockheed "Neptune"	2 Wright R-3350-26WA 2,500 hp	76'10"	100'
303.	R4Q-1	Transport, 2 engine, 5 crew, 42 troops, 35 litters, land, twin-boom tail, high-wing monoplane.	1950	Fairchild "Packet"	2 Pratt & Whitney R-4360-20W 3,500 hp	86'6"	109'3"

	Designation	Type	Assigned	Manufacturer and Name	Engine Type & Horsepower	Dimensions Length	Span
304.	R5D-2	Transport, 4 engine, 6 crew, 30-50 passenger, cargo, land, low-wing monoplane.	1950	Douglas "Skymaster"	4 Pratt & Whitney R-2000-4 1,450 hp	93'11"	117'6"
305.	TBM-3Q	Torpedo-bomber, 1 engine, 3 crew, land, mid-wing monoplane	1950	General Motors "Avenger"	Wright R-2600-20 1,900 hp	41'	54'2"
306.	TO-2	Trainer, jet, 1 engine, 2 crew, land, low-wing monoplane.	1950	Lockheed "Shooting Star"	Allison J-33 4,600 lbs. thrust	34'6"	38'10"
307.	AD-1	Attack, 1 engine, 1 crew, land/carrier, low-wing monoplane.	1951	Douglas "Skyraider"	Wright R-3350-24W 2,500 hp	39'3"	50'
308.	AD-2	Attack, 1 engine, 1 crew, land/carrier, low-wing monoplane	1951	Douglas "Skyraider"	Wright R-3350-26W 3,020 hp	39'3"	50'
309.	AD-2Q	Attack, 1 engine, 2 crew, land/carrier, low-wing monoplane.	1951	Douglas "Skyraider"	Wright R-3350-26W 3,020 hp	39'3"	50'
310.	AD-4W	Attack, 1 engine, 3 crew, land/carrier, low-wing monoplane.	1951	Douglas "Skyraider"	Wright R-3350-25WA 3,020 hp	39'3"	50'

311.	F2H-2P	Fighter, jet, 2 engine, 1 crew, land, low-wing monoplane.	1951	McDonnell "Banshee"	2 Westinghouse J-34-WE34 3,000 lbs. thrust	40'2"	41'7"
312.	F3D-1	Fighter, jet, 2 engine, 2 crew, land, high-wing monoplane.	1951	Douglas "Skyknight"	2 Westinghouse J-34-WE-32 3,250 lbs. thrust	40'2"	41'7"
313.	F3D-2	Fighter, jet, 2 engine, 2 crew, land, high-wing monoplane.	1951	Douglas "Skyknight"	2 Westinghouse J-34-WE-36 3,400 lbs. thrust.	40'2"	41'7"
314.	F4U-5NL	Fighter, 1 engine, 1 crew, land, inverted gull-wing monoplane.	1951	Vought "Corsair"	Pratt & Whitney R-2800-32W 2,000 hp	34'6"	42'
315.	F8F-2	Fighter,1 engine, 1 crew, land, low-wing monoplane.	1951	Grumman "Bearcat"	Pratt & Whitney R-2800-34W 2,300 hp	27'6"	35'6"
316.	F9F-2P	Fighter, jet, 1 engine, 1 crew, land, low-wing monoplane.	1951	Grumman "Panther"	Pratt & Whitney J42-P-8 5,700 lbs. thrust	40'	38'
317.	F9F-4	Fighter, jet, 1 engine, 1 crew, land, low-wing monoplane.	1951	Grumman "Panther"	Pratt & Whitney J42-P-8 5700 lbs. thrust	40'	38'
318.	HO4S-1	Helicopter, observation, 1 engine, 2 crew, 8 troop, land, 3 blade rotor.	1951	Sikorsky	Pratt & Whitney R-1340 550 hp	41'8"	Rotor 53'

	Designation	Type	Assigned	Manufacturer and Name	Engine Type & Horsepower	Dimensions Length	Span
319	HRS-1	Helicopter, transport, 1 engine, 2 crew, 8 troop, land, 3 blade rotor.	1951	Sikorsky	Pratt & Whitney R-1340 550 hp	41'8"	Rotor 53'
320.	HTL-3	Helicopter, trainer, 1 engine, 2 crew, land, 2 blade rotor.	1951	Bell	Franklin 6V4-178-1332 178 hp	41'4"	Rotor 35'
321.	HTL-4	Helicopter, trainer, 1 engine, 2 crew, land, 2 blade rotor.	1951	Bell	Franklin 6V4-178-1332 178 hp	41'4"	Rotor 35'
322.	OE-1 (O-1B)	Observation, 1 engine, 2 crew, land, high-wing monoplane.	1951	Cessna "Bird Dog"	Continental 0470-11 213 hp	25'	36'
323.	TBM-3R	Torpedo-bomber, 1 engine, 1 crew, 7 passenger/cargo, land, mid-wing monoplane.	1951	General Motors "Avenger"	Wright R-2600-20 1,600 hp	41'	54'2"
324.	AD-3	Attack, 1 engine, 1 crew, land/carrier, low-wing monoplane.	1951	Douglas "Skyraider"	Wright R-3350-26W 2,500 hp	39'3"	50'
325.	AD-3N	Attack, 1 engine, 1 crew, land/carrier, low-wing monoplane.	1951	Douglas "Skyraider"	Wright R-3350-26W 2,500 hp	39'3"	50'

326.	AD-4	Attack, 1 engine, 1 crew, land/carrier, low-wing monoplane.	1951	Douglas "Skyraider"	Wright R-3350-26WA 2,700 hp	39'3"	50'
327.	AD-4B	Attack, 1 engine, 1 crew, land/carrier, low-wing monoplane.	1951	Douglas "Skyraider"	Wright R-3350-26WA 2,700 hp	39'3"	50'
328.	AD-4N	Attack, 1 engine, 1 crew, land/carrier, low-wing monoplane.	1951	Douglas "Skyraider"	Wright R-3350-26WA 2,700 hp	39'3"	50'
329.	AD-4NL	Attack, 1 engine, 1 crew, land/carrier, low-wing monoplane.	1951	Douglas "Skyraider"	Wright R-3350-26WA 2,700 hp	39'3"	50'
330.	AU-1	Attack, 1 engine, 1 crew, land, inverted gull-wing monoplane.	1952	Vought "Corsair"	Pratt & Whitney R-2800-83W 2,300 hp	34'1"	41'
331.	F2H-2B	Fighter, jet, 2 engine, 1 crew, land, low-wing monoplane.	1952	McDonnell "Banshee"	2 Westinghouse J34-WE-34 3,250 lbs. thrust	40'2"	41'7"
332.	F9F-5	Fighter, jet, 1 engine, 1 crew, land/carrier, low-wing monoplane.	1952	Grumman "Panther"	Pratt & Whitney J48-P-6 7,000 lbs. thrust	42'	38'
333.	F9F-5P	Fighter, jet, 1 engine, 1 crew, land/carrier, low-wing monoplane.	1952	Grumman "Panther"	Pratt & Whitney J48-P-6 7,000 lbs. thrust	42'	38'

	Designation	Type	Assigned	Manufacturer and Name	Engine Type & Horsepower	Dimensions Length	Span
334.	HO5S-1	Helicopter, observation, 1 engine, 2 crew, 1 passenger, land, 3 blade rotor.	1952	Sikorsky	Franklin O-425-1 245 hp	27'5"	Rotor 33'
335.	HTK-1	Helicopter, trainer, 1 engine, 2 crew, land, twin interlocking 2 blade rotor.	1952	Kaman	Lycoming O-435-4 245 hp	20'6"	Rotor 41'
336.	HRS-2	Helicopter, transport 1 engine, 2 crew, 8 troops, 3 blade rotor.	1952	Sikorsky	Pratt & Whitney R-1340 550 hp	42'2"	Rotor 53'
337.	HUP-1	Helicopter, utility, 1 engine, 2-3 crew, land, twin 3 blade rotor.	1952	Piasecki "Retriever"	Continental R-975-34 525 hp	32'	Rotor 35'
338.	PV-2T	Patrol, 2 engine, 5 crew, low-wing monoplane. F7F-3N Radar trainer.	1952	Lockheed "Harpoon"	2 Pratt & Whitney R-2800-31 2,000 hp	52'	65'6"
339.	R4D-8 (C-117D)	Transport, 2 engine, 3 crew, cargo/troops, land, low-wing transport.	1952	Douglas "Skytrain"	2 Pratt & Whitney R-1820-80 1,475 Bhp	67'9"	90'
340.	TBM-3N	Torpedo-bomber, 1 engine, 3 crew, land, mid-wing monoplane.	1952	General Motors "Avenger"	Wright R-2600-20 1,600 hp	41'	54'2"

341.	TV-2 (T-33B)	Trainer, jet, 1 engine, 2 crew, land, low-wing monoplane.	1952	Lockheed "Shooting Star"	Allison J33-A-35 4,600 lbs. thrust	37'8"	38'10"
342.	AD-4L	Attack, 1 engine, 1 crew, land, low-wing monoplane.	1953	Douglas "Skyraider"	Wright R-3350-26WA 2,700 hp	39'2"	50'
343.	AD-4NA	Attack, 1 engine, 1 crew, land, low-wing monoplane.	1953	Douglas "Skyraider"	Wright R-3350-26WA 2,700 hp	39'2"	50'
344.	AD-4Q	Attack, 1 engine, 2 crew, land, low-wing monoplane.	1953	Douglas "Skyraider"	Wright R-3350-26WA 2,700 hp	39'2"	50'
345.	AD-5 (A-1E)	Attack, 1 engine, 2 crew, land, low-wing monoplane.	1953	Douglas "Skyraider"	Wright R-3350-26WA 2,700 hp	41'1"	50'
346.	F2H-4 (F-2D)	Fighter, jet, 2 engine, 1 crew, land, low-wing monoplane.	1953	McDonnell "Banshee"	2 Westinghouse J34-WE-38 3,600 lbs. thrust	40'2"	41'7"
347.	F9F-6P	Fighter, jet, 1 engine, 1 crew, land/carrier, swept-wing monoplane.	1953	Grumman "Cougar"	Pratt & Whitney J48-P-8 7,250 lbs. thrust	41'2"	34'6"
348.	HRS-3 (CH-19E)	Helicopter, transport, 1 engine, 1 crew, 8 troop, land, 3 blade rotor.	1953	Sikorsky	Wright R-1300-3 550 hp	42'2"	Rotor 53'
349.	PV-2	Patrol, 2 engine, 5 crew, land, mid-wing monoplane.	1953	Lockheed "Harpoon"	2 Pratt & Whitney R-2800-31 2,000 hp	52'	74'11"

	Designation	Type	Assigned	Manufacturer and Name	Engine Type & Horsepower	Dimensions Length	Span
350.	R4D-6Z (VC-47J)	Transport, 2 engine, 3 crew, 15-17 passenger, land, low-wing monoplane.	1953	Douglas "Skytrain:	2 Pratt & Whitney R-1830-90C 1,200 hp	64'6"	95'
351.	R4Q-2 (C-119F)	Transport, 2 engine, 5 crew, 42 troop, land, twin-boom tail, high-wing monoplane.	1953	Fairchild "Packet"	2 Wright R-3350-85 3,500 hp	86'3"	109'3"
352.	SNB-5P (RC-45J)	Trainer, 2 engine, 5 crew, land, low-wing monoplane.	1953	Beech "Navigator"	2 Pratt & Whitney R-985-50 450 hp	34'3"	47'8"
353.	AD-5N	Attack, 1 engine, 3 crew, land, low-wing monoplane	1953	Douglas "Skyraider"	Wright R-3350-26WA 2,700 hp	39'	50'9"
354.	AD-5W	Attack, 1 engine, 3 crew, land, low-wing monoplane	1954	Douglas "Skyraider"	Wright R-3350-26WA 2,700 hp	39'	50'9"
355.	F3D-2M (MF-10B)	Fighter Jet, 2 engine, 2 crew, land, high-wing monoplane.	1954	Douglas "Skyknight"	2 Westinghouse J34-WE-36 3,400 lbs. thrust	45'	50'
356.	F7U-3	Fighter, jet, 2 engine, 1 crew, land, delta wing.	1954	Vought "Cutlass"	2 Westinghouse J46-WE-8A 4,600 lbs. thrust	44'3"	39'8"

357.	F9F-6 (F-9F)	Fighter, jet, 1 engine, 1 crew, low swept-wing.	1954	Grumman "Cougar"	Pratt & Whitney J48-P-8 7,250 lbs. thrust	40'2"	34'6"
358.	FJ-2	Fighter, jet, 1 engine, 1 crew, low swept-wing.	1954	North American "Fury"	General Electric J47-GE-2 6,000 s.t.	37'6"	37'1"
359.	HTL-5 (TH-13N)	Helicopter, trainer, 1 engine, 2 crew, 2 blade rotor.	1954	Bell	Franklin 6V4-200-C-32 200 hp	41'3"	Rotor 35'2"
360	HUP-2	Helicopter, utility, 1 engine,2-3 crew, twin 3 blade rotror	1954	Piasecki "Retriever	Continental R-975-42 550 hp	43'	Rotor 35'
361.	R5D-4R (C-54R)	Transport, 4 engine, 6 crew, 30-50 passenger/cargo, low-wing monoplane.	1954	Douglas "Skymaster"	4 Pratt & Whitney R-2000-9 1,450 hp	93'11"	117'6"
362.	SNB-4	Trainer, 2 engine, 2 crew, 4 p passenger, low-wing monoplane, remanufactured SNB-1	1954	Beech "Navigator"	2 Pratt & Whitney R-985-AN-3 450 hp	34'3"	47'8"
363.	AD-6 (A-1H)	Attack-bomber, 1 engine, 1 crew, low-wing monoplane.	1955	Douglas "Skyraider"	Wright R-3350-26WA 2,500 hp	39'	50'9"
364.	FJ-3 (F-1C)	Fighter, 1 engine, 1 crew, low-swept wing.	1955	North American "Fury"	Wright J65-W-2 7,200 lbs. thrust	37'6"	38'1'

	Designation	Type	Assigned	Manufacturer and Name	Engine Type & Horsepower	Dimensions Length	Span
365.	F9F-8 (F-9J)	Fighter, 1 engine, 1 crew, low-swept wing monoplane.	1955	Grumman "Cougar"	Pratt & Whitney J48-P-8 7,250 lbs. thrust	48'7"	34'6"
366.	HOK-1 (OH-43D)	Helicopter, utility observation 1 engine 2 crew, 2 interlocking 2 blade rotors.	1955	Kaman	Pratt & Whitney R-1340-48 600 hp	25'	Rotor 47'
367.	OE-2 (O-1C)	Observation, 1 engine, 2 crew, high-wing monoplane.	1955	Cessna "Bird Dog"	Continental O-470-2 265 hp	25'8"	36'
368.	R4Y-1 (C-131F)	Transport, 2 engine, 4 crew, low-wing monoplane.	1955	Convair	2 Pratt & Whitney R-2800 2,000 hp	79'2"	105'4"
369.	PV-2T1	Patrol, 2 engine, 5 crew, low-wing monoplane. F3D-2 Radar trainer.	1956	Lockheed "Harpoon"	2 Pratt & Whitney R-2800-31 2,000 hp	52'	65'6"
370.	FJ-3M (MF-1C)	Fighter, 1 engine, 1 crew, low-swept wing monoplane	1956	North American "Fury"	Wright J65-W-4B-18 7,800 lbs. thrust	37'6"	37'1"
371.	FJ-4 (F-1E)	Fighter, 1 engine, 1 crew, low-swept wing monoplane.	1956	North American "Fury"	Wright J65-W-18 7,800 lbs. thrust	39'1"	37'6"

372.	F3D-2T	Night-fighter, 2 engine, 2 crew, high-wing monoplane. F4D-1 Radar trainer.	1956	Douglas "Skyknight"	2 Westinghouse J34-WE-36 4,000 lbs. thrust each	45'	50'
373.	F3D-2Q (EF-10B)	Night-fighter, 2 engine, 2 crew, high-wing monoplane.	1956	Douglas "Skyknight"	2 Westinghouse J34-WE-36 4,000 lbs. thrust each	45'	50'
374.	F2H-3 (F-2C)	All-weather fighter, 2 engine, 1 crew, low-wing monoplane.	1956	McDonnell "Banshee"	2 Westinghouse J34-WE-34 3,600 lbs. thrust each	40'2"	41'8"
375.	F9F-8B (AF-9J)	Fighter, 1 engine, 1 crew, low-swept wing monoplane.	1956	Grumman "Cougar"	Pratt & Whitney J48-P-8 7,250 lbs. thrust	41'7"	34'6"
376.	F9F-8P (RF-9J)	Fighter, 1 engine, 1 crew, low-swept wing monoplane.	1956	Grumman "Cougar"	Pratt & Whitney J48-P-8 7250 lbs. thrust	41'7"	34'6"
377.	F4D-1 (F-6A)	Fighter, 1 engine, 1 crew, delta wing.	1956	Douglas "Skyray"	Pratt & Whitney J57-P-8 10,000 lbs. thrust, 14,500 lbs. w/afterburner	45'5"	33'6"
378.	HR2S-1 (CH-37C)	Helicopter, assault transport heavy, 3 crew, 36 passenger, 5 blade rotor.	1956	Sikorsky "Mojave"	2 Pratt & Whitney R-2800-54 , 1,900 hp	82'10"	Rotor 72'

	Designation	Type	Assigned	Manufacturer and Name	Engine Type & Horsepower	Dimensions Length	Span
379.	T-28B	Trainer, 1 engine, 2 crew, low-wing monoplane.	1956	North American "Trojan"	Wright R-1820-86 1450 hp	34'4"	40'7"
380.	T-34B	Trainer, 1 engine, 2 crew, low-wing monoplane.	1957	Beech "Mentor"	Continental O-470-13 225 hp	25'11"	32'10"
381.	A4D-1 (A-4A)	Attack, 1 engine, 1 crew, low delta type monoplane.	1957	Douglas "Skyhawk"	Wright J65-W-4 7200 lbs. thrust	39'1"	27'6"
382.	A4D-2 (A-4B)	Attack, 1 engine, 1 crew, low delta type monoplane.	1957	Douglas "Skyhawk"	Wright J65-W-16A 7700 lbs. thrust	39'1"	27'6"
383.	FJ-4B (AF-1E)	Fighter, 1 engine, 1 crew, low-swept wing monoplane.	1957	North American "Fury"	Wright J65-W-16A 7700 lbs. thrust	37'6"	39'1"
384.	F9F-8T (TF-9J)	Fighter, 1 engine, 2 crew, low-swept wing monoplane.	1957	Grumman "Cougar"	Pratt & Whitney J48-P-8A 7250 s.t.	44'5"	34'6"
385.	HUS-1 (UH-34D)	Helicopter, utility transport, 1 engine, 3 crew, 12 passenger, 4 blade rotor.	1957	Sikorsky "Seahorse"	Wright R-1820-84 1525 hp	47'2"	Rotor 56'
386.	HUS-1A (UH-34E)	Helicopter, utility transport, 1 engine, 3 crew, 12 passenger, amphibious pontoons, 4 blade rotor.	1957	Sikorsky "Seahorse"	Wright R-1820-84 1525 hp	47'2"	Rotor 56'

387.	U-1A	Utility liaison transport, 1 engine, 2 crew, 12 passenger, high-wing monoplane.	1957	DeHavilland "Otter"	Pratt & Whitney R-1340-AN-3 600 hp	41'10"	58'
388.	F8U-1 (F-8A)	Fighter, 1 engine, 1 crew, high-swept wing monoplane.	1958	Chance Vought "Crusader"	Pratt & Whitney J57-P-4A 16,000 lbs. thrust w/afterburner	54'3"	35"8"
389.	F8U-1P (RF-8A)	Fighter, 1 engine, 1 crew, high-swept wing monoplane.	1958	Chance Vought "Crusader"	Pratt & Whitney J57-P-4A 16,000 lbs. thrust w/afterburner	54'3"	35'8"
390.	F8U-1E (F-8B)	Fighter, 1 engine, 1 crew, high-swept wing monoplane.	1959	Chance Vought "Crusader"	Pratt & Whitney J57-P-12 16,000 lbs. thrust w/afterburner	54'3"	35'8"
391.	F8U-2 (F-8C)	Fighter, all weather, 1 engine, 1 crew, w/tailfins, high-swept wing monoplane.	1959	Chance Vought "Crusader"	Pratt & Whitney J57-P-16 16,000 lbs. thrust w/afterburner	54'3"	35'8"
392.	A4D-2N (A-4C)	Attack, all weather, 1 engine, 1 crew, low-delta type monoplane.	1960	Douglas "Skyhawk"	Wright J65-W-4 7200 lbs. thrust	39'1"	27'6"
393.	UO-1 (U-11A)	Transport, utility, 2 engine, 2 crew, 4 place, low-wing monoplane.	1960	Piper "Aztec"	2 Lycoming O-540A 250 hp	27'6"	37'

	Designation	Type	Assigned	Manufacturer and Name	Engine Type & Horsepower	Dimensions Length	Span
394.	HUS-1Z (VH-34D)	Helicopter, utility transport, 1 engine, 3 crew, 12 passenger, 4 blade rotor.	1960	Sikorsky "Seahorse"	Wright R-1820-84 1525 hp	47'2"	Rotor 56'
395.	UF-1 (HU-16C)	Utility transport, 2 engine, 5 crew, 10-22 passenger, high-wing monoplane, amphibian.	1960	Grumman "Albatross"	2 Wright R-1820-76A 1425 hp	62'10"	96'8"
396.	F8U-2N (F-8D)	Fighter, all weather, intercepter, 1 engine, 1 crew, high-swept wing monoplane.	1960	Chance Vought "Crusaider"	Pratt & Whitney J57-P-4A 16,000 lbs. thrust w/afterburner	54'3"	35'2"
397.	R5D-5R (C-54T)	Transport, 4 engine, 6 crew, 30-50 passenger, cargo, low-wing monoplane.	1960	Douglas "Skymaster"	4 Pratt & Whitney R-2000-11 1450 hp	93'11"	117'6"
398.	HSS-2 (SH-3A)	Helicopter, transport 2 engine, 3 crew, 26 passenger, 4 blade rotor.	1961	Sikorsky "Sea Bat"	General Electric T58-GE-8E shaft turbine, 1250 hp	54'9"	Rotor 62'
399.	R4D-8Z (VC-117D)	Transport, 2 engine, 3 crew, low-wing monoplane.	1961	Douglas "Skytrain"	2 Wright R-1820-80 1200 hp	67'8"	90'
400.	GV-1 (KC-130F)	Transport, tanker, 4 engine 5-7 crew, 92 passenger, high-wing monoplane.	1961	Lockheed "Hercules"	4 Allison T56-A7 turboprop, 4,050 hp	97'9"	132'7"

401.	T2V-1 (T-1A)	Trainer, 1 engine, jet, 2 crew, low-wing monoplane.	1961	Lockheed "Seastar"	Allison J33-A-22 6100 lbs. thrust	38'	42'
402.	F4H-1 (F-4B)	Fighter, all weather interceptor, 2 engine, 2 crew, low-wing monoplane.	1962	McDonnell "Phantom II"	General Electric J79-GE-8 17,000 lbs. thrust	58'3"	38'5"
403.	HSS-2Z (VH-3A)	Helicopter, transport, 2 engine, 3 crew, 12 passenger, 4 blade rotor.	1962	Sikorsky "Seahorse"	General Electric T58-GE-8B shaft turbine 1250 hp	54'9"	Rotor 62'
404	F8U-2NE (F-8E)	Fighter, all weather interceptor, 1 engine, 1 crew, high-swept wing monoplane.	1962	Ling-Temco-Vought "Crusader"	Pratt & Whitney J57-P-4A 16,000 lbs. thrust w/afterburner	54'6'	35'2"
405.	A4D-5 (A-4E)	Attack, 1 engine, 1 crew, low-delta type wing monoplane.	1964	Douglas "Skyhawk"	Pratt & Whitney J52-P-6 8500 lbs. thrust	42'11"	27'5"
406.	A-6A	Attack, all weather, 2 engine, 2 crew, swept mid-wing monoplane.	1964	Grumman "Intruder"	2 Pratt & Whitney J52-P-6 8500 lbs. thrust	54'7"	53'
407.	CH-46A	Helicopter, transport, 2 engine, 3 crew, 17-25 passenger, 4000 lbs. cargo, dual 3 blade rotors.	1964	Boeing-Vertol "Sea Knight"	General Electric T58-GE-8B shaft turbine, 1250 hp	45'7" Rotors folded	Rotors 50'

	Designation	Type	Assigned	Manufacturer and Name	Engine Type & Horsepower	Dimensions Length	Span
408.	UH-1E	Helicopter, assault support, 1 engine, 2 crew, 7 passengers, 2 blade rotor.	1964	Bell "Iroquois"	Lycoming T53-L-11 shaft turbine, 1,100 hp	41'8"	Rotor 48'
409.	UH-43C	Helicopter, utility, 1 engine, 2 crew, 10 passenger, twin interlocking 2 interlocking 2 blade rotors.	1964	Kaman "Huskie"	Lycoming T53-L-1B shaft turbine, 825 bhp	25'2"	Rotor 47'
410.	UH-2A	Helicopter, utility, 1 engine, 2 crew, 11 passenger, 4 blade rotor.	1964	Kaman "Seasprite"	General Electric T58-GE-8B shaft turbine, 1250 hp	52'2"	Rotor 44'
411.	RF-4B	Reconnaissance, 2 engine, 2 crew, low-swept wing monoplane.	1965	McDonnell "Phantom II"	2 General Electric J79-GE-8, 16,000 lbs. thrust	63'	38'5"
412.	EA-6A	Electronic warfare aircraft, 2 engine, 2 crew, mid swept-wing monoplane.	1965	Grumman "Intruder"	2 Pratt & Whitney J52-P-6 8500 lbs. thrust	55'6"	53'
413.	TA-4E	Advanced jet trainer, 1 engine, 2 crew, low-delta type wing, monoplane.	1965	Douglas "Skyhawk"	Pratt & Whitney J52-P-8A 9,300 lb	42"6"	27'5"

414.	CH-53A	Helicopter, cargo/troop transport, 2 engine, 3 crew, 6 blade rotor.	1965	Sikorsky "Sea Stallion"	2 T64-GE-6 2,850 shp	56'6" Rotor Stowed	Rotor 72'
415.	S-2E	ASW aircraft, 2 engine, 4 crew, high-wing monoplane, used with ASW gear removed.	1965	Grumman "Tracker"	2 Curtiss Wright R-1820-82A 1,525 bht	43'6"	72'7"
416.	CH-46D	Helicopter, transport, 2 engine, 3 crew, 17-25 passenger 4,000 lbs. cargo, dual 3 blade rotors	1966	Boeing "Sea Knight"	2 General Electric T58-GE-10 1,400 hp	84'4"	Rotor 51'
417.	TA-4F	Advance jet trainer version of the A-4E, 1 engine, 2 crew, low delta type wing monoplane.	1966	McDonnell Douglas "Skyhawk"	1 Pratt & Whitney J52-P-8A 9,300 lb	42'5"	27'5"
418.	US-2B	2 engine, 4 crew, high-wing monoplane, modified for utility duties.	1966	Grumman "Tracker"	2 Wright R-1820-82 1,425 Bhp	42'	69'8"
419.	F-4J	Fighter, 2 engine, 2 crew, low-wing monoplane.	1967	McDonnell "Phantom II"	2 General Electric J79-GE-1U 17,859 lb	58'3"	38'4"
420.	F-8H	Fighter, 1 engine, 1 crew, swept wing, remanufactured F-8D.	1967	Vought "Crusaider"	Pratt & Whitney J-57-420 19,000 lb	54'3"	35'8"

	Designation	Type	Assigned	Manufacturer and Name	Engine Type & Horsepower	Dimensions Length	Span
421.	CH-46F	Helicopter, transport, 2 engine, 3 crew, 17-25 passenger, 4,000 lbs. cargo, dual 3 blade rotors.	1968	Boeing "Sea Knight"	General Electric T58-GE-10 1,400 hp	83'4""	Rotor 50'
422.	OV-10A	Visual armed reconnaissance, 2 engine, turbo prop, 2 crew, high-wing monoplane.	1968	North American "Bronco"	2 Garrett-AiResearch T-76-G-10/12 715 Shp	41'6"	40'
423.	TC-4C	An A-6A Bombardier/Navigator trainer, 2 engine, 3 crew.	1968	Grumman "Gulfstream"	2 Rolls Royce Dart Mark 529-8X 1,910 Shp	67'9"	90'
424.	TC-117D	Transport, 2 engine, 3 crew, 33 passenger, low wing monoplane.	1968	Douglas "Skytrain"	2 Wright R-1820-80 1,435 hp	67'9"	90'
425.	US-2A	Search type modified for utility duty, 2 engine, 4 crew, high-wing monoplane.	1968	Grumman "Tracker"	2 Wright R-1820-82 1,425 hp	42'	69'8"
426.	OV-10D	Observation, 2 engine 2 crew, twin tail, high-wing monoplane.	1968	Rockwell "Bronco"	2 Garrett T76-G-416/417 715 hp	41'7"	40'

427.	UH-2B	Helicopter, 2 engine, 2 crew, 4 blade rotor.	1969	Kaman "Seasprite"	2 General Electric T58-GE-8 1,3350 shp	52'7"	Rotor 44'
428.	CH-53D	Assault transport helicopter, 2 engine, 3 crew, main 6 blade rotor and tail pylon fold for storage aboard ships.	1969	Sikorsky "Sea Stallion"	2 General Electric T64-GE-413 3,230 shp	56'6"	Rotor 72'3"
429.	AH-1G	Attack helicopter, 1 engine, 2 crew in tandem.	1969	Bell "Cobra"	Lycoming T53-L-13 1,400 shp	44'7"	Rotor 44'
430.	T-28C	Advanced trainer, 1 engine, 2 crew, low-wing monoplane.	1970	North American "Trojan"	1 Wright R-1820-26 1,275 bhp	34' 5"	40' 7"
431.	HH-1K	Based on the UH-1E as a search and rescue aircraft with improved avionics, 2 blade rotor.	1970	Bell "Iroquois"	Lycoming T-53-L-13 1,400 SHP	42'7"	Rotor 44'
432.	C-131G	Transport, 2 engine, -- crew, 44 passenger, low-wing monoplane.	1970	Convair "Convair Liner"	2 Pratt & Whitney R-2800-52W 2,200 bhp	79'2"	105'4"
433.	TA-4J	Light attack trainer, 1 engine, 2 crew, low delta wing.	1970	Douglas "Skyhawk"	1 Pratt & Whitney J52-P-6 8,500lb	40'4"	27'6"

	Designation	Type	Assigned	Manufacturer and Name	Engine Type & Horsepower	Dimensions Length	Span
434.	KA-6D	Configured with refueling hose and reel, 2 engine, 2 crew	1971	Grumman "Intruder"	2 Pratt & Whitney J52-P-8B 9,300 lb	54'9"	53'
435.	A-4M	Light attack, 1 engine, 1 crew, low delta wing monoplane.	1971	McDonnell Douglas "Skyhawk"	1 Pratt & Whitney J-52-P408 9,900 lb	41'3"	27'5"
436.	AV-8A	Light attack aircraft with vertical takeoff and landing capability, 1 engine, 1 crew high swept wing monoplane.	1971	Hawker Siddeley "Harrier"	1 Rolls Royce Bristol Pegasus F402-RR-401 1,300 lb	45' 5"	25' 2"
437.	AH-1J	Armed tactical helicopter, 1 engine, 2 crew in tandem, 2 blade rotor.	1971	Bell "Huey Twin Cobra"	1 United Aircraft of Canada (P & W) T400-CP-400 1,134 SHP	53' 4"	44' 0"
438.	UH-1N	Helicopter, utility and reconnaissance, 2 engine, 2 crew in tandem, 2 blade rotor.	1971	Bell "Iroquois"	1 United Aircraft of Canada (P & W) T400-CF-400 1,530 SHP	57'3"	Rotor 48'
439.	T-39D	Trainer for inflight training in radar scope interpretation, 2 engine, -- crew, low swept wing monoplane.	1972	North American "Saberliner"	2 Pratt and Whitney J60-P-3A 2,570 lb	44' 0"	44' 5"

440.	SH-3G	Primary mission as ASW. Also provides search and rescue missions, 5 blade rotor.	1972	Sikorsky "Sea King"	2 General Electric T-58-GE-10 1,400 Shp	72'7"	Rotor 62'
441.	C-118B	Long range cargo/personnel transport, 4 engine, --crew, low-wing monoplane.	1973	Douglas "Liftmaster"	4 Pratt & Whitney R-2800-52W 1,900 bhp	107'	117'6"
442.	F-4N	Rebuilt F-4B upgrade	1973	McDonnell "Phantom II"	2 General Electric J79-GE-8 17,000 lb	58'3"	38'5"
443.	A-6E	All weather, low altitude attack, 2 engine, 2 crew, mid swept wing monoplane.	1974	Grumman "Intruder"	2 Pratt & Whitney J52-P-8A/B 9,300 lb	54" 9"	52' 0"
444.	CT-39G	Transport, 2 engine, 3 crew, 7 passenger, low wing monoplane.	1974	Rockwell "Saberliner"	2 Pratt & Whitney JT12A-8 3,300 lb	43'9"	44'5"
445.	HH-46A	Helicopter, search and rescue, 2 engine, 3 crew, twin 3 blade rotors.	1974	Boeing Vertol "Sea Knight"	2 General Electric T-58-GE-8B 1,250 hp	83'4"	Rotor 50'
446.	C-9B	General purpose transport, 2 engine, 5 crew, 90 passenger. low-wing monoplane.	1975	McDonnell Douglas "Skytrain II"	2 Pratt & Whitney JT8D-9 14,500 hp	119'3"	93'3"

	Designation	Type	Assigned	Manufacturer and Name	Engine Type & Horsepower	Dimensions Length	Span
447.	VH-3D	Helicopter, modified as an executive aircraft for the President of the United States. 5 blade rotor.	1975	Sikorsky "Sea King"	2 General Electric T58-GE-10 1,400 shp	72'8"	Rotor 62'
448.	T-2C	Trainer, 2 engine, 2 crew in tandem, mid-wing monoplane.	1975	Rockwell International "Buckeye"	2 General Electric J85-GE-4 2,950 lb	38'3"	38'
449.	TAV-8A	Trainer, vertical take-off, 1 engine, 2 crew in tandem, high swept wing monoplane.	1976	McDonnell Douglas/British Aerospace "Harrier"	1 Rolls Royce F402-RR-406A 21,450 lb	55'11"	25'4"
450.	CH-46E	Helicopter, transport, 2 engine, 4 crew, 4,000 lbs. cargo, 3 blade rotor.	1976	Boeing Vertol "Sea Knight"	2 General Electric T58-GE-16 24,300 lb	84'4"	Rotor 51'
451.	VH-1N	UH-1N helicopters converted to staff configuration. 2 blade rotor.	1976	Bell "Iroquois""	1 United Aircraft of Canada (P & W) T400-CF-400 1,530 SHP	57'3"	Rotor 48'
452.	KC-130R	Combination transport and inflight refueler, 4 engine, 5 crew, high-wing monoplane.	1976	Lockheed "Hercules"	4 Allison T56-A-16 4,200 shp	99' 5"	132' 5"

453.	F-4S	Fighter, 2 engine, 2 crew in tandem, low-wing monoplane.	1978	McDonnell "Phantom II"	2 General Electric J79-GE-108 11,100 lb	58" 3"	38' 4"
454.	AH-1T	Attack helicopter, 2 engine, 2 crew in tandem, 2 blade rotor.	1978	Bell "Sea Cobra"	2 Pratt & Whitney T-400-WV-402 1,970 Shp.	58'	Rotor 48'
455.	EA-6B	All weather, electronic warfare aircraft, 2 engine, 4 crew, mid swept-wing monoplane.	1978	Grumman "Prowler"	2 Pratt & Whitney J52-P-8A 9,300 lb	59' 4"	53' 0"
456.	OA-4M	Observation,1 engine,1 crew, low delta wing monoplane.	1979	McDonnell Douglas "Skyhawk"	1 Pratt & Whitney J52-P-8A 9,300 lb	42'5"	27'5"
457.	UC-12B	Transport, cargo or 13 passenger, 2 engine, 2 crew, low-wing monoplane.	1980	Beech "Huron"	2 Pratt & Whitney PT6A-41 850 shp	43'9"	54'6"
458.	VC-118B	Transport, 4 engine, 6 crew, executive interior, low-wing monoplane.	1980	Douglas "Liftmaster"	4 Pratt & Whitney R-2800-52W, 2,500 hp	107'	117'6"
459.	AV-8C	Vertical take-off, 1 engine, 1 crew, high swept wing monoplane.	1981	McDonnell Douglas "Harrier"	1 Rolls Royce Bristol Pegasus F402-RR-401 1,300 lb.	45'5"	25'2"

	Designation	Type	Assigned	Manufacturer and Name	Engine Type & Horsepower	Dimensions Length	Span
460.	CH-53E	Helicopter, 3 engine, 3 crew, a refueling probe, 7 blade rotor.	1981	Sikorsky "Super Stallion"	3 General Electric T64-GE-416A 4,380 shp	99'	Rotor 79'
461.	AV-8B	Vertical take-off, 1 engine, 1 crew, high swept wing monoplane.	1983	McDonnell Douglas "Harrier II"	1 Rolls-Royce F402-RR-406A 2 1,450 lb	46'4"	30'4"
462.	FA-18A	Fighter, 2 engine, 1 crew, shoulder-wing monoplane.	1983	McDonnell Douglas "Hornet"	2 General Electric F404-GE-400 16,000 lb	56'	36'6"
463.	AH-1W	Helicopter, 2 engine, 2 crew in tandem, 2 blade rotor.	1986	Bell	2 General Electric T700-GE-401 3,380 shp	44'7"	Rotor 44'
464.	VH-60N	Helicopter, 2 engine, 3 crew, 11 passengers, single rotor	1987	Sikorsky "Sea Hawk"	2 General Electric T700-GE-401 1,690 shp	64'5"	Rotor 53'8"
465.	TAV-8B	Trainer version of AV-8B, 2 crew, 1 engine, vertical take off, high wing monoplane.	1987	McDonnell Douglas "Harrier II"	1 Rolls-Royce F402-RR-406 14,540 lb	50'4"	30'3"
466.	T34C	Trainer, 1 engine, 2 crew in tandum, low wing monoplane.	1991	Beech Aircraft Corp. "Mentor"	1 Pratt & Whitney PT6A-25 715 shp	28'8"	33'5"

467.	C-20G	Transport, 2 engine, 4 crew, 26 passenger/cargo, low wing monoplane.	1998	Gulfstream Aerospace "Gulfstream II"	2 Rolls-Royce MK 511-8 11,400 lb	77'10"	83'1"
468.	UC-35C	Transport, 2 engine, 2 crew, 12 passenger, low wing, monoplane.	1999	Cessna "Citation V"	2 Pratt & Whitney JT15D-5D	48'11"	52'2"
469.	MV-22	Tiltrotor, vertical take off, 3 crew, 2 engine.	2000	Bell-Boeing "Osprey"	2 Allison Gas Turbine T406-AD-400 6,150 shp	57'3"	Rotor 38'
470.	KC-130J	Combination transport and inflight refueler, 4 engine, 5 crew, high wing monoplane.	2000	Lockheed "Hercules"	4 Allison T56-A- 5,300 shp	97'9"	132'7"

Made in the USA
Monee, IL
05 November 2026

67385029R10050